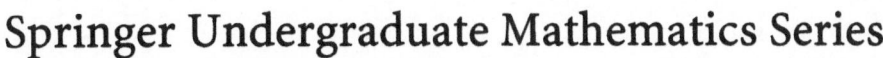

Springer Undergraduate Mathematics Series

Springer-Verlag London Ltd.

Other books in this series

D.L. Johnson

Symmetries

With 60 Figures

 Springer

D.L. Johnson, BSc, MSc, PhD
Department of Mathematics, University of Nottingham, University Park,
Nottingham NG7 2RD, UK

British Library Cataloguing in Publication Data
Johnson, D.L. (David Lawrence), 1943-
    Symmetries. - (Springer undergraduate mathematics series)
    1. Symmetry
    I. Title
    516.1'5
ISBN 978-1-85233-270-9

Library of Congress Cataloging-in-Publication Data
Johnson, D.L.
    Symmetries / D.L. Johnson
        p.   cm. -- (Springer undergraduate mathematics series, ISSN 1615-2085)
    Includes bibliographical references and index.
    Additional material to this book can be downloaded from http://extra.springer.com.
    ISBN 978-1-85233-270-9   ISBN 978-1-4471-0243-4 (eBook)
    DOI 10.1007/978-1-4471-0243-4
    1. Symmetry groups. I. Title. II. Series.
QA174.2.J64   2001
512'.2—dc21   2001020163

Springer Undergraduate Mathematics Series ISSN 1615-2085
ISBN 978-1-85233-270-9

http://www.springer.co.uk

Typesetting: Camera ready by the author, Kate MacDougall and Thomas Unger

12/3830-54321 Printed on acid-free paper   SPIN 10901533

For Mrs. H

Preface

"... many eminent scholars, endowed with great geometric talent, make a point of never disclosing the simple and direct ideas that guided them, subordinating their elegant results to abstract general theories which often have no application outside the particular case in question. Geometry was becoming a study of algebraic, differential or partial differential equations, thus losing all the charm that comes from its being an art."

H. Lebesgue, *Leçons sur les Constructions Géométriques*, Gauthier-Villars, Paris, 1949.

This book is based on lecture courses given to final-year students at the University of Nottingham and to M.Sc. students at the University of the West Indies in an attempt to reverse the process of expurgation of the geometry component from the mathematics curricula of universities. This erosion is in sharp contrast to the situation in research mathematics, where the ideas and methods of geometry enjoy ever-increasing influence and importance. In the other direction, more modern ideas have made a forceful and beneficial impact on the geometry of the ancients in many areas. Thus trigonometry has vastly clarified our concept of angle, calculus has revolutionised the study of plane curves, and group theory has become the language of symmetry.

To illustrate this last point at a fundamental level, consider the notion of congruence in plane geometry: two triangles are congruent if one can be moved onto the other so that they coincide exactly. This property is guaranteed by each of the familiar conditions SSS, SAS, SAA and RHS. So congruent triangles are just copies of the same triangle appearing in (possibly) different places. This makes it clear that congruence is an equivalence relation, whose three defining properties correspond to properties of the moves mentioned above:

- reflexivity – the identity move,

- symmetry – inverse moves,

- transitivity – composition of moves.

Thinking of these moves as transformations, for which the associative law holds automatically, we have precisely the four axioms for a group: closure, associativity, identity and inverses.

The group just described underlies and in a sense determines plane geometry. It is called the Euclidean group and occupies a dominant position in this book. Its elements are isometries, as defined in Chapter 1, and a detailed study of these occupies Chapters 2 and 4. The rather bulky Chapters 3 and 5 are intended as crash courses on the theory of groups and group presentations respectively, and both lay emphasis on groups that are semidirect products. Such groups arise in the classification of discrete subgroups of the Euclidean group in Chapters 6, 7 and 8, and corresponding tessellations (or tilings) appear in Chapter 9. Regular tessellations of the sphere are classified in Chapter 10, and tessellations of other spaces, such as the hyperbolic plane, form the subject of Chapter 11. Finally, the notions of polygon in 2-space and polyhedron in 3-space are generalised in Chapter 12 to that of a polytope in n-dimensional Euclidean space. Regular polytopes are then defined using group theory and classified in all dimensions. The classification contains some surprises in dimension 4 and is achieved by as elegant a piece of mathematics as you might imagine.

The exercises at the end of each chapter form an integral part of the book, being designed to reinforce your grasp of the material. A large majority are more or less routine, but a handful of more challenging problems are included for good measure. Solutions to most of them, or at least generous hints, are given later, and suggestions for background, alternative and further reading appear towards the end of the book.

It is a pleasure to acknowledge my gratitude to a number of people: to J.A. Green, B.H. Neumann, J.H. Conway and R.C. Lyndon for influence and guidance over the years, and likewise to John Humphreys, Bob Laxton, Jim Wiegold and Geoff Smith for valuable encouragement; to Maxine Francis, Kate MacDougall and Aaron Wilson for skilful preparation of the typescript and diagrams; to all at Springer-Verlag, especially David Anderson, Nick Wilson, Susan Hezlet, David Ireland and Karen Borthwick, for efficient handling of matters connected with production; and last but not least to the students who provided much useful feedback on my lectures.

D.L.J.

Contents

1
Metric Spaces and their Groups

In many of the physical sciences a fundamental role is played by the concept of length: units of length are used to measure the distance between two points. In mathematics the idea of distance, as a function that assigns a real number to a given pair of points in some space, is formalised in terms of a few reasonable-looking properties, or axioms, and the result is called a metric on that space. Having defined a structure such as this on a set, it is natural to study those transformations, or maps, of such sets which preserve that structure. The requirement that these maps be invertible then leads naturally into the theory of groups.

Many types of groups arise in this way. Important examples are permutation groups, linear groups, Galois groups and symmetry groups. The story of the last of these begins as follows.

1.1 Metric Spaces

Our intuitive conception of distance is made precise in the following definition.

Definition 1.1

A **metric** on a set X is a map $d: X \times X \to \mathbb{R}$ with the following properties:

M1) $d(x, y) \geq 0 \ \forall x, y \in X$, with equality if and only if $x = y$;

M2) $d(x, y) = d(y, x) \; \forall x, y \in X$;

M3) $d(x, y) + d(y, z) \geq d(x, z) \; \forall x, y, z \in X$.

A set X with a metric d is called a **metric space**, written (X, d).

Each of these axioms is in accordance with our intuition. Thus, referring to elements of X as points, M1 says that the distance between two distinct points is positive, and the distance from a point to itself is zero. M2 says that d is symmetric: the distance between two points is the same in either direction. M3 is the famous **triangle inequality**: the direct route between two points is the shortest. Time for some examples.

Example 1.1

Let X be any set and for $x, y \in X$ define $d(x, y) = 0$ if $x = y$, $d(x, y) = 1$ if $x \neq y$.

To check the axioms, observe that all three parts of M1 are trivial consequences of the definition of d, and the same goes for M2. As to M3, the triangle inequality can only fail if the right-hand side is 1 and both terms on the left are equal to zero. But this means that

$$x \neq z, \quad x = y \quad \text{and} \quad y = z,$$

and this is a contradiction. Hence, M3 holds and d is indeed a metric on X, sometimes referred to as the **discrete metric**.

Example 1.2

Let $X = \mathbb{R}$ and for $x, y \in \mathbb{R}$ define $d(x, y) = |x - y|$, where the modulus $|x|$ of a real number x is defined in the usual way:

$$|x| = \begin{cases} x & \text{if } x \geq 0, \\ -x & \text{if } x < 0. \end{cases}$$

The axioms correspond to obvious properties of the modulus function, and their verification, as in the next example, is left as an exercise. Both of the next two examples represent attempts to generalise this example from the real line $\mathbb{R}$ to the Cartesian plane $\mathbb{R}^2$.

Example 1.3

Let $X = \mathbb{R}^2$ and for $\mathbf{x} = (x_1, x_2)$, $\mathbf{y} = (y_1, y_2) \in \mathbb{R}$ define

$$d(\mathbf{x}, \mathbf{y}) = |x_1 - y_1| + |x_2 - y_2|.$$

This metric is sometimes referred to as the **Manhattan metric** on $\mathbb{R}^2$.

Example 1.4

Again let $X = \mathbb{R}^2$ and for $\mathbf{x} = (x_1, x_2)$, $\mathbf{y} = (y_1, y_2) \in \mathbb{R}^2$ define

$$d(\mathbf{x}, \mathbf{y}) = +\sqrt{(y_1 - x_1)^2 + (y_2 - x_2)^2}\,, \tag{1.1}$$

the non-negative root. Unwieldy though it looks, this is the one we want. It is called the **Pythagorean metric**, and when we refer to $\mathbb{R}^2$ as a metric space it is this metric we have in mind.

As usual, M1 and M2 are pretty obvious, being consequences of simple facts about real numbers such as

$$x^2 = 0 \iff x = 0, \quad x^2 = (-x)^2.$$

M3 on the other hand is equivalent to the assertion that the length of one side of a plane triangle is less than or equal to the sum of the lengths of the other two, whence its epithet. For a formal proof in terms of coordinates, take any three points

$$\mathbf{x} = (x_1, x_2), \quad \mathbf{y} = (y_1, y_2), \quad \mathbf{z} = (z_1, z_2)$$

in the plane, and define real numbers a_1, a_2, b_1, b_2 by setting

$$(y_1, y_2) = (x_1 + a_1, x_2 + a_2),$$
$$(z_1, z_2) = (y_1 + b_1, y_2 + b_2)$$
$$= (x_1 + a_1 + b_1, x_2 + a_2 + b_2).$$

Then it is required to prove that

$$\sqrt{a_1^2 + a_2^2} + \sqrt{b_1^2 + b_2^2} \geq \sqrt{(a_1 + b_1)^2 + (a_2 + b_2)^2}\,. \tag{1.2}$$

Since squares of real numbers are non-negative, we have

$$(a_1 b_2 - a_2 b_1)^2 \geq 0,$$

which implies that

$$a_1^2 b_2^2 + a_2^2 b_1^2 \geq 2 a_1 b_1 a_2 b_2.$$

Adding $a_1^2 b_1^2 + a_2^2 b_2^2$ to both sides and taking square roots,

$$\sqrt{a_1^2 + a_2^2}\,\sqrt{b_1^2 + b_2^2} \geq a_1 b_1 + a_2 b_2.$$

Multiplying by 2, adding $a_1^2 + a_2^2 + b_1^2 + b_2^2$ to both sides and again taking square roots, we obtain (1.2).

1.2 Isometries

Definition 1.2

An **isometry** of a metric space (X, d) is a bijective map $u \colon X \to X$ that preserves distance:

$$d(xu, yu) = d(x, y) \quad \forall x, y \in X. \tag{1.3}$$

The set of all isometries of (X, d) is denoted by $\mathrm{Isom}(X, d)$, or just $\mathrm{Isom}(X)$ where d is taken for granted.

Notice that maps are written on the right: the image of x under u is denoted by xu rather than $u(x)$. This is standard practice in those parts of mathematics where the main interest in maps is centred on their composition, for then uv denotes the composite of two maps u and v in this order (first u, then v).

Another point to note is that this definition involves some redundancy. A bijection is by definition a map that is both injective and surjective. It is left to the Exercises to show that any map $u \colon X \to X$ satisfying (1.3) is necessarily injective (but not necessarily surjective).

Now recall that bijective maps are precisely those that have inverses: there is a map u^{-1} such that

$$uu^{-1} = 1 = u^{-1}u, \tag{1.4}$$

where 1 denotes the identity map, $x1 = x \ \forall x \in X$. Then, assuming that u satisfies (1.3), we compute

$$d(xu^{-1}, yu^{-1}) = d(xu^{-1}u, yu^{-1}u) = d(x1, y1) = d(x, y)$$

for all $x, y \in X$. This shows that if u is an isometry of X, then so is u^{-1}.

Next observe that the identity map $1 \colon X \to X$ is obviously an isometry and that composition of maps is always associative, that is, independent of the bracketing.

Finally, consider the composite uv of two isometries $u, v \in \mathrm{Isom}(X)$. Since uv has an inverse, namely $v^{-1}u^{-1}$, it is a bijection. And since v, u satisfy (1.3), we have

$$d(xuv, yuv) = d(xu, yu) = d(x, y)$$

for all $x, y \in X$. Hence, uv is again an isometry of X. We say that $\mathrm{Isom}(X)$ is closed under composition of maps.

The last three paragraphs prompt the next definition and constitute a proof of our first theorem.

Definition 1.3

A **group** is a set closed under an associative binary operation and containing an identity and the inverse of each of its elements.

Theorem 1.1

The set Isom(X) *of isometries of a metric space* X *forms a group under composition of maps.*   □

Let us say a few words about the group Isom(X) in each of the four examples in the previous section.

In the case of the discrete metric on a set X, it is not hard to show that any bijection $u: X \to X$ satisfies (1.3). The group of all such bijections, or **permutations** of x, is called the **symmetric group** on X. When X is a finite set, say $X = \{1, 2, \ldots, n\}$, with n elements, this group is often referred to as the symmetric group of **degree** n and denoted by S_n. The number of its elements is $n!$.

We shall study the important example of the real line with metric

$$d(x, y) = |x - y|, \quad x, y \in \mathbb{R}, \tag{1.5}$$

in rather more detail. A complete description of the group Isom$(\mathbb{R}, d)$ will be obtained in the next section, where many of the ideas needed later in the book will be introduced.

From our point of view, the Manhattan metric on $\mathbb{R}^2$ is useful mainly as a source of examples and exercises. These can be compared and contrasted with corresponding properties and aspects of the fourth example.

The group Isom$(\mathbb{R}^2, d)$, with d given by (1.1), forms the chief object of study in this book. It is called the **Euclidean group** and denoted by $\mathbb{E}$. As a kind of warming-up exercise, we shall now, as promised, present a detailed description of its little brother.

1.3 Isometries of the Real Line

For the duration of this section and the next, let G denote the group Isom$(\mathbb{R}, d)$ of isometries of the real line with the usual metric, that given by (1.5). We note at the outset the value of the group concept in describing its elements and analysing its structure.

Let u be an arbitrary element of G and suppose that u maps 0 to c, $0u = c \in \mathbb{R}$. Let $t = t(c)$ denote the map that adds the fixed number c to any element of $\mathbb{R}$,

$$t: \mathbb{R} \to \mathbb{R}, \quad x \mapsto x + c. \tag{1.6}$$

It is not hard to show that t is an isometry of $\mathbb{R}$, so $t \in G$. Then its inverse $t^{-1} \in G$ and also the composite $v = ut^{-1} \in G$. Since t maps 0 to c, t^{-1} must map c back to 0, and thus

$$0v = 0ut^{-1} = ct^{-1} = 0,$$

that is, v fixes 0. Then for any $x \in \mathbb{R}$, $x \neq 0$,

$$|x| = d(0, x) = d(0v, xv) = d(0, xv) = |xv|,$$

so that there are only two possibilities for $xv : xv = \pm x$. In particular, $1v = (-1)^\varepsilon$, where $\varepsilon = 0$ or 1, and we have two cases.

Case 0: $\varepsilon = 0$. So far we have

$$0v = 0, \quad 1v = 1, \quad xv = \pm x$$

for any $x \in \mathbb{R}$. We claim that $xv = x$ for all $x \in \mathbb{R}$. If not, there is some $x \neq 0$ with $xv = -x$, and then

$$|x - 1| = d(1, x) = d(1v, xv) = d(1, -x) = |x + 1|.$$

Then $x - 1 = \pm(x + 1)$ and in either case we get a contradiction: $-1 = +1$ or $x = 0$, respectively. So there is no such x, the claim is valid, and v is the identity map, $v = ut^{-1} = 1$. This means that $u = t$, and we have identified u in this case.

Case 1: $\varepsilon = 1$. Now we have

$$0v = 0, \quad 1v = -1, \quad xv = \pm x$$

for any $x \in \mathbb{R}$. As in Case 0, we establish the claim that $xv = -x$ for all $x \in \mathbb{R}$, and v is equal to the map

$$r: \mathbb{R} \to \mathbb{R}, \quad x \mapsto -x, \tag{1.7}$$

which is clearly an isometry. So $v = ut^{-1} = r$, whence $u = rt$, and we have again identified u, this time as the map

$$rt: \mathbb{R} \to \mathbb{R}, \quad x \mapsto -x + c. \tag{1.8}$$

We have shown that an arbitrary isometry u of $\mathbb{R}$ admits a factorisation

$$u = r^\varepsilon t(c), \quad \varepsilon = 0 \text{ or } 1, \ c \in \mathbb{R}, \tag{1.9}$$

where $t(c)$ and r are given by (1.6) and (1.7). Because this factorisation is unique (exercise), we refer to it as a **normal form** for the elements of G. To complete the description of G, it remains to calculate the product (or composite) of two such forms,

$$u = r^\varepsilon t(c), \quad v = r^\delta t(d), \tag{1.10}$$

where $\varepsilon, \delta \in \{0, 1\}$ and $c, d \in \mathbb{R}$.

To put the right-hand side of the equation

$$uv = r^\varepsilon t(c) r^\delta t(d)$$

into normal form, first take the case when $\delta = 0$. It is clear from (1.6) that

$$t(c)t(d) = t(c + d), \quad c, d \in \mathbb{R}, \tag{1.11}$$

and so the normal form is $uv = r^\varepsilon t(c + d)$ in this case. When $\delta = 1$, the second occurrence of r in the product is in the wrong place. To correct this, observe that $r^2 = (rt)^2 = 1$ from (1.7) and (1.8), so that $tr = r^{-1}t^{-1} = rt(-c)$, and the normal form is $uv = r^{\varepsilon+1}t(d - c)$ in this case.

Combining these two cases, we can state that the multiplication in G is given by the formula

$$uv = r^{\varepsilon+\delta}t(d + (-1)^\delta c), \tag{1.12}$$

where u, v are given by (1.10). Note that since $r^2 = 1$, ε and δ are treated as integers modulo 2 $(1 + 1 = 0)$, so that the right-hand side of (1.12) is indeed in normal form and the algebraic description of the group G is complete.

The above analysis of the structure of G contains a number of points that need to be emphasised and stored up for future reference. So many in fact that we devote a whole new section to listing them.

1.4 Matters Arising

We retain all the notation used in the previous section, and begin with the alternative $\varepsilon = 0$ or 1 in the normal form (1.9). This defines a fundamental and far-reaching dichotomy. Some points of contrast between the resulting two types of isometry are described below.

First take the case $\varepsilon = 0$, when $u = t(c)$ is defined by (1.6). When $c = 0$ this is the identity map, $t(0) = 1$. When $c \neq 0$, $t(c)$ moves every point $x \in \mathbb{R}$ the same distance $|c|$, to the right or left according as c is positive or negative. So it makes sense to refer to such a map as a **translation**. Some characteristic properties of translations $t = t(c)$ are as follows:

1. t is **order-preserving** (OP): $x < y \Rightarrow xt < yt$,

2. t has no **fixed points**: $xt \neq x \ \forall x \in \mathbb{R}$, unless $c = 0$,

3. t is **aperiodic**: $t^n \neq 1 \ \forall n \in \mathbb{N}$, unless $c = 0$.

Another important property of translations follows from (1.11): the subset

$$T = \{t(c) \mid c \in \mathbb{R}\}$$

of G forms a group. It is reasonable to call this a **subgroup** of G.

Now let $\varepsilon = 1$, so that $u = rt(c)$ is defined by (1.8). Notice that the arithmetic mean of any $x \in \mathbb{R}$ and its image $xu = -x + c$ takes the constant value $c/2$. Thus u moves every point $x \in \mathbb{R}$ to the point equidistant from and on the other side of the point $c/2$. So it makes sense to refer to such a map as a **reflection**, and write $u = rt(c) = r(c/2)$. Some characteristic properties of reflections $r = r(c/2)$ are as follows:

1. r is **order-reversing** (OR): $x < y \Rightarrow xr > yr$,

2. r has a **fixed point**: $xr = x$ when $x = c/2$,

3. r is **periodic**: $r^2 = 1$.

Another point of contrast is that the set

$$\{rt(c) \mid c \in \mathbb{R}\} = rT, \quad \text{where } r = r(0),$$

of reflections does not form a group. It has the obvious properties

$$T \cap rT = \emptyset, \quad T \cup rT = G$$

and is called a **left coset** of the subgroup T in G.

In deference to the properties numbered 1 of the subsets T and rT of G, we sometimes write

$$T = G^+, \quad rT = G^-.$$

Despite all the points of difference between G^+ and G^-, there is one property they have in common: they have the same number of elements. In fact, the maps

$$\mathbb{R} \longrightarrow G^+, \quad G^+ \longrightarrow G^-$$
$$c \longmapsto t(c), \quad t(c) \longmapsto rt(c)$$

are both bijections, and so

$$|G^+| = |G^-| = |\mathbb{R}|,$$

the cardinal of the continuum.

Another point of difference between translations and reflections illustrates an important group-theoretical concept. For all $c, d \in \mathbb{R}$ we have

$$t(c)t(d) = t(c + d) = t(d)t(c),$$

whereas, except when $c = d$,

$$r(c)r(d) = t(2(d - c)) \neq t(2(c - d)) = r(d)r(c).$$

Thus, translations commute and reflections (in general) do not. A group in which any two elements commute is said to be **abelian**. So T is abelian and G is non-abelian.

It turns out that we know rather a lot about abelian groups. In fact there is a complete classification in the finitely generated case, and this will be described later. Non-abelian groups on the other hand are by no means classified, even in the finite case. Two ways of measuring non-abelianness are provided by the simple operations of conjugation and commutation, defined as follows.

Let x and y be two elements in an arbitrary group. Then the **conjugate** of x by y is defined by the equation

$$x^y = y^{-1}xy,$$

and the **commutator** of x and y is given by

$$[x, y] = x^{-1}y^{-1}xy.$$

The connection with commutativity is provided by the obvious equivalence

$$xy = yx \Leftrightarrow x^y = x \Leftrightarrow [x, y] = 1.$$

These operations have a number of useful properties and lead to two important constructions, **conjugacy classes** and the **commutator subgroup**, respectively, that will play a significant role in what follows. We content ourselves for the moment with a brief look at the effect of conjugation in the group $G = \text{Isom}(\mathbb{R})$.

The conjugates of any translation $t(c) \in G^+$ and any reflection $r(c) \in G^-$ by an arbitrary isometry $u \in G$ are given by

$$u^{-1}t(c)u = t((-1)^\varepsilon c), \quad u^{-1}r(c)u = r(cu), \tag{1.13}$$

where $\varepsilon = 0$ or 1 according as u is OP or OR and cu is the image under u of $c \in \mathbb{R}$. These equations illustrate the general point that the effect of conjugation on an isometry is to produce an isometry of the same kind.

A special case of the first equation in (1.13) is the rule

$$r^{-1}t(c)r = t(-c), \tag{1.14}$$

where $c \in \mathbb{R}$ and $r = r(0)$, which was used in the derivation of the formula (1.12). Indeed the multiplication in G is completely determined by the relations (1.11) and (1.14) and the fact that $r^2 = 1$. More about this later.

1.5 Symmetry Groups

Definition 1.4

Let X be a metric space and F a subset of X. Then by a **symmetry** of F we mean an isometry of X that fixes F (as a set). The set of all symmetries of F is denoted by $\mathrm{Sym}(F)$. In symbols,

$$\mathrm{Sym}(F) = \{u \in \mathrm{Isom}(X) \mid Fu = F\}.$$

It should be emphasised here that the equation $Fu = F$ does not assert that u fixes every element of F, but merely that F and $Fu = \{xu \mid x \in F\}$ are equal as sets. Thus, any isometry u of X that fixes F must also fix its complement $X \setminus F$, and so $\mathrm{Sym}(F) = \mathrm{Sym}(X \setminus F)$ for any $F \subseteq X$.

Now we know from Theorem 1.1 that $\mathrm{Isom}(X)$ forms a group under composition of maps, and it is a matter of routine to check that the same is true for $\mathrm{Sym}(F)$ for any $F \subseteq X$:

$$F1 = F, \quad F = Fu \Rightarrow Fu^{-1} = Fuu^{-1} = F1 = F,$$
$$Fu = F = Fv \Rightarrow Fuv = Fv = F,$$

and we have the following theorem.

Theorem 1.2

For any subset F of a metric space X, $\mathrm{Sym}(F)$ is a subgroup of $\mathrm{Isom}(X)$. $\square$

Let us pursue the example $G = \mathrm{Isom}(\mathbb{R})$ studied in Section 1.3, taking F to be the set $\mathbb{Z}$ of integers. We write $H = \mathrm{Sym}(\mathbb{Z})$ for the group of isometries of $\mathbb{R}$ that fix $\mathbb{Z}$ as a set. Then any $u \in H$ must map 0 to an integer, $0u = n \in \mathbb{Z}$ say. At the same time, we know from (1.9) that $u = r^{\varepsilon}t(c)$, where $r = r(0)$, $\varepsilon = 0$ or 1, $c \in \mathbb{R}$. Since $r^2 = 1$, we have $r^{\varepsilon}u = t(c)$ and so

$$c = 0t(c) = 0r^{\varepsilon}u = 0u = n,$$

whence u has the form $r^{\varepsilon}t(n)$, where $n \in \mathbb{Z}$. But any such isometry maps any integer m to

$$mr^{\varepsilon}t(n) = ((-1)^{\varepsilon}m)t(n) = n + (-1)^{\varepsilon}m,$$

which is again in $\mathbb{Z}$. It follows that the elements of H are precisely those isometries of the form $r^{\varepsilon}t(n)$, $n \in \mathbb{Z}$:

$$H = \{r^{\varepsilon}t(n) \mid n \in \mathbb{Z}, \varepsilon = 0 \text{ or } 1\}.$$

Letting $t = t(1)$, an easy induction based on (1.11) shows that $t(n) = t^n$ for all $n \in \mathbb{N}$, and even for all $n \in \mathbb{Z}$, since

$$t(n)t(-n) = t(0) = 1 = t^0 = t^n t^{-n}.$$

Thus, every $u \in H$ can be expressed in terms of just two elements, $r = r(0)$, $t = t(1)$:

$$u = r^\varepsilon t^n, \quad \varepsilon = 0 \text{ or } 1, \, n \in \mathbb{Z}, \tag{1.15}$$

and we say that r and t **generate** H and write $H = \langle r, t \rangle$. In terms of these generators, multiplication in H is given by the relations

$$t^m t^n = t^{m+n}, \quad r^2 = 1, \quad r^{-1} t^n r = t^{-n},$$

$m, n \in \mathbb{Z}$, and all of these are consequences of just two:

$$r^2 = 1, \quad r^{-1} t r = t^{-1},$$

which we therefore call **defining relations** for this group. Hence, the entire structure of H is encoded in the **presentation**:

$$H = \langle r, t \mid r^2 = 1, r^{-1} t r = t^{-1} \rangle. \tag{1.16}$$

This group is called the **infinite dihedral group** and denoted by D_∞.

Sadly, there is no such presentation for the group $G = \text{Isom}(\mathbb{R})$. Indeed a finitely generated group can have at most countably many elements, and G has the cardinal of the continuum. A rather more subtle fundamental difference between G and its subgroup H, and one that is more important for us, is arrived at geometrically as follows. Consider the set xH of images of a fixed real number x under elements of H:

$$xH = \{xu \mid u \in H\},$$

sometimes called the **orbit** of x under H. It follows from (1.15) that

$$xH = \{n \pm x \mid n \in \mathbb{Z}\},$$

and we ask the following question. How close can two elements of this set be to one another?

A little thought shows that the answer is 1 when x is an integer or half-integer, and twice the distance from x to the nearest integer or half-integer otherwise. In every case there is a positive minimum distance μ, so that *the interval $(x - \mu, x + \mu)$ contains no point of xH other than x itself.* We express this property of H, which is obviously not shared by G, by saying that H is a **discrete** (or, more properly, discontinuous) subgroup of G.

EXERCISES

1.1. Prove that the rule $d(x, y) = |x - y|$ defines a metric on the set $\mathbb{R}$ of real numbers.

1.2. The same for the Manhattan metric on $\mathbb{R}^2$.

1.3. Show that any distance-preserving map on a metric space is necessarily injective.

1.4. Find a counter-example to show that a distance-preserving map on a metric space need not be surjective.

1.5. Prove that any permutation of a set preserves the discrete metric.

1.6. Show that the symmetric group S_n has $n!$ elements.

1.7. Try to extend the types of isometry of $\mathbb{R}$ described in Section 1.3 to transformations of $\mathbb{R}^2$ that preserve a) the Manhattan metric, b) the Pythagorean metric. Can you find a transformation that preserves the latter but not the former?

1.8. Prove that the translations $t(c)$ and the reflection r are isometries of $\mathbb{R}$.

1.9. Prove the uniqueness of the form (1.9) for isometries of $\mathbb{R}$.

1.10. Write down the inverse of the isometry $rt(c)$ of $\mathbb{R}$ in normal form.

1.11. Show from first principles that no non-trivial isometry of $\mathbb{R}$ can fix two distinct points. Deduce that an isometry of $\mathbb{R}$ is determined by its effect on two distinct points.

1.12. Prove that there are exactly three possibilities for the number of distinct powers of an isometry u of $\mathbb{R}$.

1.13. Find a formula for the composite $r(c)r(d)$ of two reflections of $\mathbb{R}$.

1.14. Check that $(x^y)^z = x^{(yz)}$ for any three elements of any group. Deduce that conjugacy is an equivalence relation on any group.

1.15. Describe the partition into classes of the group $G = \text{Isom}(\mathbb{R})$ under conjugacy.

1.16. Calculate the commutators $[u, v]$ of various pairs of elements $u, v \in G = \text{Isom}(\mathbb{R})$. Find a general form for these commutators and so identify the subgroup G' of G which they generate.

1.17. Verify the formulae (1.13) for conjugation in $G = \text{Isom}(\mathbb{R})$.

1.18. Why is it not possible to replace the first of the two formulae in (1.13) by the more consistent-looking equation $u^{-1}t(c)u = t(cu)$?

1.19. Describe the subgroups $\mathrm{Sym}(\mathbb{Q})$ and $\mathrm{Sym}(\pm 1)$ of $\mathrm{Isom}(\mathbb{R})$, where $\mathbb{Q}$ denotes the set of rational numbers and ± 1 the two-point set $\{-1, 1\}$. Is either of these subgroups discrete?

1.20. Prove that the infinite dihedral group D_∞ given by (1.15) is generated by the two elements $x = r$, $y = rt$. Write down enough relations on x, y to define the multiplication in this group, and so obtain an alternative presentation for D_∞.

Isometries of the Plane

The plane here means the ordinary Euclidean plane, parametrised à la Descartes by ordered pairs (x, y) of real numbers and hence denoted by $\mathbb{R}^2$. According to the previous chapter, an isometry of $\mathbb{R}^2$ is a distance-preserving bijection from $\mathbb{R}^2$ to itself, where distance is defined by the usual (Pythagorean) metric. Having discussed various general properties of isometries of $\mathbb{R}^2$ in Section 2.1, we turn in Section 2.2 to a description of the three basic types, translations, rotations and reflections, and their special properties. It turns out that every isometry of $\mathbb{R}^2$ can be expressed in a canonical way in terms of these three. This is the substance of the important and illuminating normal form theorem which occupies Section 2.3. To complete the picture of the group of isometries of $\mathbb{R}^2$ it remains to describe in normal form the product of two normal forms. This is done in Section 2.4, where conjugation of isometries plays a role whose significance will emerge in later chapters.

2.1 Congruent Triangles

Everybody knows what a triangle is. Since two points in the plane lie on a unique line, three non-collinear points, say $A, B, C \in \mathbb{R}^2$, determine three distinct lines. Each of these lines divides $\mathbb{R}^2$ into two half-planes, and so together the lines partition the plane into eight regions. Well, seven actually, as one of them is empty. Six of the seven are unbounded, and the seventh is what we call the **triangle** $\Delta = ABC$. Δ has three **sides**, AB, BC, CA, and three (interior)

angles, $B\widehat{C}A$, $C\widehat{A}B$, $A\widehat{B}C$. The points A, B, C are called the **vertices** of Δ, and their disposition determines an **orientation** of Δ, which we declare to be positive if they occur consecutively in this order anticlockwise on the boundary of Δ, and negative otherwise. Triangles have a lot of interesting properties.

The fundamental question of when two triangles, say Δ and Δ', shall be regarded as the same is answered as follows. Δ and Δ' are said to be **congruent** if their vertices can be labelled, say by A, B, C and A', B', C' respectively, in such a way that corresponding sides are of equal length and corresponding angles are of equal magnitude:

$$|AB| = |A'B'|, \quad |BC| = |B'C'|, \quad |CA| = |C'A'|, \tag{2.1}$$

$$B\widehat{C}A = B'\widehat{C}'A', \quad C\widehat{A}B = C'\widehat{A}'B', \quad A\widehat{B}C = A'\widehat{B}'C'. \tag{2.2}$$

Note that in this definition angles, like lengths, $|AB| := d(A, B)$, are taken to be non-negative. A basic result of plane geometry, which we glibly assume, asserts that the six equations in (2.1) and (2.2) involve redundancy. For example, any two of Eq. (2.2) imply the third, since the angle-sum of a triangle takes the constant value π. In fact, any one of the following four familiar conditions

$$\textbf{SSS}, \quad \textbf{SAS}, \quad \textbf{SAA}, \quad \textbf{RHS}$$

is sufficient for congruence (but **SSA**, **AAA** are not). The first of these, that two triangles with corresponding sides equal are congruent, is the one we need.

Theorem 2.1

Any isometry u of $\mathbb{R}^2$

(i) *maps any triangle ABC to a congruent triangle,*

(ii) *preserves angles, and*

(iii) *maps lines to lines.*

Proof

(i) Since u is an isometry, (2.1) holds with $A' = Au$, $B' = Bu$, $C' = Cu$. Hence ABC, $A'B'C'$ are congruent (SSS).

(ii) Any angle $\alpha < \pi$ can be realised as the angle $C\widehat{A}B$ in some triangle ABC. Then, using the notation and the result of part (i), $C'\widehat{A}'B' = C\widehat{A}B = \alpha$ too. If $\alpha > \pi$ take an exterior angle, and if $\alpha = \pi$ use the law of trichotomy.

(iii) Let l be any line in $\mathbb{R}^2$, fix distinct points B, C on l, and let l' denote the line through $B' = Bu$, $C' = Cu$. Then, for a point $A \in \mathbb{R}^2$, it follows from part

(ii) that

$$Au \notin lu \Leftrightarrow A \notin l \Leftrightarrow 0 < C\hat{A}B < \pi$$
$$\Leftrightarrow 0 < C'\hat{A}uB' < \pi \Leftrightarrow Au \notin l',$$

whence $lu = l'$, as required. □

We now proceed to give an important application of Theorem 1.1, whence we know that the set $\text{Isom}(\mathbb{R}^2)$ forms a group under composition of maps. We call this the **Euclidean group** and denote it by $\mathbb{E}$.

Theorem 2.2

An isometry of $\mathbb{R}^2$ is determined by its effect on any three non-collinear points.

Proof

Let O, P, Q be any three non-collinear points of $\mathbb{R}^2$ and let $u_1, u_2 \in \mathbb{E}$ have the same effect on each:

$$Ou_1 = Ou_2, \quad Pu_1 = Pu_2, \quad Qu_1 = Qu_2. \tag{2.3}$$

Then we have to prove that u_1 and u_2 are equal as maps, that is,

$$Ru_1 = Ru_2 \quad \text{for every point } R \in \mathbb{R}^2. \tag{2.4}$$

Since $\mathbb{E}$ is a group (Theorem 1.1), we can write $u = u_1 u_2^{-1} \in \mathbb{E}$, whereupon (2.3) and (2.4) can be rewritten

$$Ou = O, \quad Pu = P, \quad Qu = Q, \tag{2.5}$$
$$Ru = R \quad \forall R \in \mathbb{R}, \tag{2.6}$$

respectively, and our task reduces to proving the implication $(2.5) \Rightarrow (2.6)$.

So assume that $u \in \mathbb{E}$ satisfies (2.5) and let $R \in \mathbb{R}^2$ be arbitrary. Since u preserves distance and $Ou = O$, it follows that R and Ru are equidistant from O, that is, Ru lies on the circle C_1 centre O radius $|OR|$. Similarly, Ru lies on the circle C_2 centre P radius $|PR|$.

So $Ru \in C_1 \cap C_2$. Since C_1, C_2 are not concentric ($O \neq P$), they intersect in at most two points. But since by construction R lies on them both, $C_1 \cap C_2$ contains at least one point. Thus $|C_1 \cap C_2| = 1$ or 2. In the former case, the point of intersection is $R = Ru$. In the latter (see Fig. 2.1), we put $C_1 \cap C_2 = \{R, R'\}$, so that the line l through O, P is the perpendicular bisector of R and R'. Since by hypothesis Q does not lie on l, we have $|QR| \neq |QR'|$. The fact that R, Ru are equidistant from Q now implies that $Ru = R$, and (2.6) holds in this case too. □

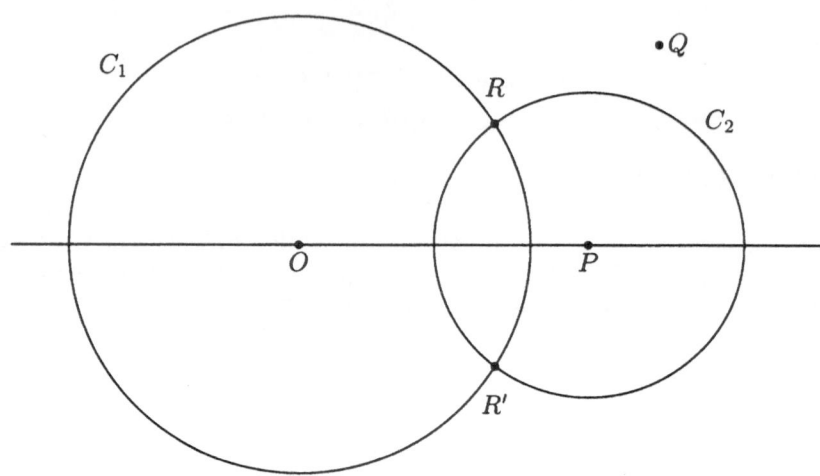

Figure 2.1   Intersection of C_1 and C_2

2.2 Isometries of Different Types

Definition 2.1

A **translation** t is a map that moves every point of the plane through a fixed distance in a fixed direction. In symbols,

$$t\colon \mathbb{R}^2 \to \mathbb{R}^2, \quad (x_1, x_2) \mapsto (x_1 + a_1, x_2 + a_2),$$

where (x_1, x_2) is an arbitrary point of $\mathbb{R}^2$ in Cartesian coordinates, and $\mathbf{a} = (a_1, a_2)$ is a constant vector. We write $t = t(\mathbf{a})$ and call the direction of $\mathbf{a}$ the **axis** of t (thought of as an equivalence class of parallel lines).

Three key properties of translations are that they are **orientation preserving** (OP), have no fixed points (except when $\mathbf{a} = \mathbf{0}$) and compose according to the rule

$$t(\mathbf{a})t(\mathbf{b}) = t(\mathbf{a} + \mathbf{b}), \tag{2.7}$$

where $\mathbf{a}, \mathbf{b} \in \mathbb{R}^2$ are arbitrary and the $+$ sign stands for componentwise (or vector) addition. It follows from this last property that the translations in $\mathbb{E}$ form a subgroup, which we denote by $\mathbb{T}$ and refer to as the **translation subgroup** of $\mathbb{E}$.

Definition 2.2

A **rotation** s is a map that moves every point of the plane through a fixed angle about a fixed point, called the **centre**. Taking the centre O to be the origin of polar coordinates (ρ, θ), we have

$$s: \mathbb{R}^2 \to \mathbb{R}^2, \quad (\rho, \theta) \mapsto (\rho, \theta + \alpha),$$

where α is a constant angle, $\alpha \in \mathbb{R} \bmod 2\pi$. We write $s = s(O, \alpha)$.

Three key properties of rotations are that they are OP, have just one fixed point (except when $\alpha = 0$) and compose according to the rule

$$s(O, \alpha)s(O, \beta) = s(O, \alpha + \beta), \tag{2.8}$$

where the $+$ sign stands for addition in $\mathbb{R} \bmod 2\pi$. It follows from this last property that the rotations with centre O form a subgroup of $\mathbb{E}$, which we denote by $\mathbb{S}_0$. The harder problem of how to compose rotations with distinct centres will be solved in Chapter 4.

Definition 2.3

A **reflection** r is a map that moves every point of the plane to its mirror-image in a fixed line. This line, l say, is called the **axis** of r and we write $r = r(l)$. Thus, given $P \in \mathbb{R}^2$, if $P \in l$ then $Pr = P$, and if $P \notin l$, Pr is the unique point of $\mathbb{R}^2$ such that l is the perpendicular bisector of P and Pr.

Three key properties of reflections are that they are **orientation reversing** (OR), fix every point on a line and satisfy

$$r(l)^2 = 1. \tag{2.9}$$

The composition of reflections with different axes is big business, and will occupy the whole of Chapter 4 below.

Focussing for a moment on the isometry $s = s(O, \pi/2)$, consider the effect of applying it twice. Two successive rotations about O through $\pi/2$ amount to a single rotation through π, and so

$$s^2 = s(O, \pi)$$

in accordance with (2.8). In Cartesian coordinates, it is clear that

$$(x, y)s^2 = (-x, -y),$$

and the effect of s^2 is to multiply by -1. The effect of s may therefore be safely construed as multiplication by $\sqrt{-1}$, and we have attached a physical meaning to this dubious quantity. If we denote it by i, and the point $(1, 0)$ by 1, we get

$$(0, 1) = (1, 0)s = 1i = i,$$

and
$$(x, y) = (1, 0)x + (0, 1)y = x + iy.$$

The points of $\mathbb{R}^2$ thus correspond naturally to complex numbers in the Argand diagram. In this representation, the effects on $z = x + iy$ of translation t by $w = a + ib$, rotation s about 0 through α, reflection r in the x-axis are

$$zt = z + w, \quad zs = ze^{i\alpha}, \quad zr = \bar{z},$$

where $\bar{z} = x - iy$ is the complex conjugate of z. Combining these, we can manufacture the isometries

$$z \mapsto ze^{i\alpha} + w, \quad z \mapsto \bar{z}e^{i\alpha} + w, \tag{2.10}$$

which are OP, OR respectively. We shall now show that, for $w \in \mathbb{C}$ and $\alpha \in \mathbb{R} \bmod 2\pi$, the isometries in (2.10) comprise a complete list of elements of $\mathbb{E}$.

2.3 The Normal Form Theorem

Theorem 2.3

Fix a point O and a line l in $\mathbb{R}^2$ with $O \in l$. Then any element $u \in \mathbb{E}$ can be written uniquely in the form

$$u = r^\varepsilon st, \tag{2.11}$$

where r denotes reflection in l, $\varepsilon = 0$ or 1, $s \in \mathbb{S}_0$ and $t \in \mathbb{T}$. The subgroup $\mathbb{E}^+$ of OP isometries consists of just those u with $\varepsilon = 0$.

Proof

To prove the existence of the form for a given $u \in \mathbb{E}$, first let t be the translation sending O to Ou. Then ut^{-1} fixes O.

Next let P be any point on l with $P \neq O$. Since

$$0 \neq d(O, P) = d(Out^{-1}, Put^{-1}) = d(O, Put^{-1}),$$

there is a rotation $s \in \mathbb{S}_0$ sending P to Put^{-1}. Then $ut^{-1}s^{-1}$ fixes P and O. Finally, let Q be any point not on l. Then the points Q and $Qut^{-1}s^{-1}$ are equidistant from both O and P. Since circles with distinct centres can meet in at most two points, it follows as in the proof of Theorem 2.2 that the points Q and $Qut^{-1}s^{-1}$ are either equal or mirror-images in l. Putting $\varepsilon = 0$ in the first case and $\varepsilon = 1$ in the second, the isometry $ut^{-1}s^{-1}r^\varepsilon$ fixes Q as well as O and

P. Since by construction O, P, Q are not collinear, it follows from Theorem 2.2 that $ut^{-1}s^{-1}r^{\varepsilon} = 1$, that is, $u = r^{\varepsilon}st$, as required.

Now observe that, since s, t are OP and r is OR, the isometry $r^{\varepsilon}st$ is OP or OR according as $\varepsilon = 0$ or 1, which proves the assertion about $\mathbb{E}^{+}$.

Finally, to prove the uniqueness of the form, assume that

$$r^{\varepsilon}st = r^{\delta}s't',$$

where $\varepsilon, \delta \in \{0, 1\}$, $s, s' \in \mathbb{S}_0$, $t, t' \in \mathbb{T}$. It follows from the previous paragraph that $\varepsilon = \delta$ and, cancelling the r if necessary, we have $st = s't'$. But then $s'^{-1}s = t't^{-1}$ belongs to $\mathbb{S}_0 \cap \mathbb{T}$ and so, being a translation that fixes a point, it must be the identity map. Thus, $s = s'$, $t = t'$, and the uniqueness of (2.11) is established. $\qquad\qquad\qquad\qquad\qquad\qquad\qquad\qquad\qquad\qquad\qquad\qquad\qquad\square$

2.4 Conjugation of Isometries

The normal form (2.11) provides a concise and unambiguous description of the elements of $\mathbb{E}$. To obtain such a description of the group structure on $\mathbb{E}$, we need to express in normal form the product of two normal forms. That is, we seek a formula

$$r^{\eta}s(\gamma)t(\mathbf{c}) = r^{\delta}s(\alpha)t(\mathbf{a}) \cdot r^{\varepsilon}s(\beta)t(\mathbf{b}), \qquad (2.12)$$

where r, s, t stand for reflection in l, rotation about O, translation respectively, and $\delta, \varepsilon, \eta \in \mathbb{Z} \bmod 2$, $\alpha, \beta, \gamma \in \mathbb{R} \bmod 2\pi$, $\mathbf{a}, \mathbf{b}, \mathbf{c} \in \mathbb{R}^2$ respectively. The problem is to express the parameters η, γ, $\mathbf{c}$ (to be determined) in terms of δ, α, $\mathbf{a}$, ε, β, $\mathbf{b}$ (assumed to be given). Four possible approaches spring immediately to mind.

Method 1 is to use the algorithm implicit in the first part of the proof of the normal form theorem. This begins by applying the left- and right-hand sides of (2.12) (call them L and R) to the point O. Noting that r and s fix O, we compute

$$\begin{aligned}
\mathbf{c} = Ot(\mathbf{c}) = OL &= OR \\
&= Ot(\mathbf{a})r^{\varepsilon}s(\beta)t(\mathbf{b}) \\
&= \mathbf{a}r^{\varepsilon}s(\beta)t(\mathbf{b}) \\
&= \mathbf{a}r^{\varepsilon}s(\beta) + \mathbf{b},
\end{aligned}$$

and we have succeeded (more or less) in expressing $\mathbf{c}$ in terms of the other parameters. After this it gets worse.

Method 2 uses the idea of conjugation and is thus naturally favoured by group theorists. Abbreviating (2.12) in an obvious way, first manipulate as follows:

$$r^\eta s'' t'' = r^\delta s' t' r^\varepsilon s t$$
$$= r^{\delta+\varepsilon} r^{-\varepsilon} (s' t') r^\varepsilon s t$$
$$= r^{\delta+\varepsilon} s'^{r^\varepsilon} t'^{r^\varepsilon} s t$$
$$= r^{\delta+\varepsilon} \cdot s'^{r^\varepsilon} s \cdot t'^{r^\varepsilon s} t. \tag{2.13}$$

Next, show that $s'^{r^\varepsilon} \in \mathbb{S}_0$ and $t'^{r^\varepsilon s} \in \mathbb{T}$ using the following general principle: given isometries u and v, if u maps P to Q, then its conjugate u^v maps Pv to Qv. Finally, deduce from the uniqueness part of Theorem 2.3 that

$$r^\eta = r^{\delta+\varepsilon}, \quad s'' = s'^{r^\varepsilon} s, \quad t'' = t'^{r^\varepsilon s} t.$$

This yields among others the fact, already obvious from considerations of orientation, that $\eta = \delta + \varepsilon \bmod 2$.

Method 3 will make a strong appeal to the clairvoyant reader. Express all the quantities involved in terms of reflections using the techniques described in Chapter 4 below, and then manipulate them in a careful and systematic way until you get the answer.

Method 4 is to use the representation of $\mathbb{E}$ on the complex plane $\mathbb{C}$ described in Section 2.2 above. It therefore violates the principle of purity of method. But then so does almost everything else in this book. Besides, it's quick and it works.

Having replaced $\mathbf{a}, \mathbf{b}, \mathbf{c} \in \mathbb{R}^2$ by the corresponding $a, b, c \in \mathbb{C}$, apply the right-hand side of (2.12) to the complex number z in polar form $\rho e^{i\theta}$, first when $\delta = 0$. If $\varepsilon = 0$ also, we get

$$zs(\alpha)t(a)s(\beta)t(b) = \rho e^{i\theta} s(\alpha)t(a)s(\beta)t(b)$$
$$= \left(\rho e^{i(\theta+\alpha)} + a\right) s(\beta)t(b)$$
$$= \rho e^{i(\theta+\alpha+\beta)} + ae^{i\beta} + b$$
$$= zs(\alpha+\beta)t(ae^{i\beta} + b).$$

Similarly, if $\varepsilon = 1$,

$$zs(\alpha)t(a)rs(\beta)t(b) = \left(\rho e^{i(\theta+\alpha)} + a\right) rs(\beta)t(b)$$
$$= \left(\rho e^{-i(\theta+\alpha)} + \bar{a}\right) s(\beta)t(b)$$
$$= \rho e^{i(\beta-\alpha-\theta)} + \bar{a}e^{i\beta} + b$$
$$= \bar{z}s(\beta-\alpha)t(\bar{a}e^{i\beta} + b).$$

Finally, if $\delta = 1$, replace the z in each last line by $\bar{z}$.

Theorem 2.4

The composition of isometries of $\mathbb{R}^2$ in normal form is given by the formula (2.12), where

$$\eta = \delta + \varepsilon, \quad \gamma = \beta + (-1)^\varepsilon \alpha, \quad \mathbf{c} = \mathbf{a} r^\varepsilon s(\beta) + \mathbf{b}. \qquad \square$$

As a bonus, the conjugates appearing in (2.13) can now be evaluated as special cases using the uniqueness part of Theorem 2.3.

Corollary 2.1

Let r be reflection in a line l, $s = s(\beta)$ rotation through $\beta \in \mathbb{R} \bmod 2\pi$ about a point $O \in l$, and $t(\mathbf{a})$ translation by a vector $\mathbf{a} \in \mathbb{R}$. Then

$$t(\mathbf{a})^r = t(\mathbf{a}r), \quad t(\mathbf{a})^s = t(\mathbf{a}s), \quad s(\beta)^r = s(-\beta). \tag{2.14}$$

$$\square$$

Some comments are appropriate. First, in view of the normal form theorem and the fact that translations commute with each other, the first two of these equations show that the translation subgroup $\mathbb{T}$ is closed under conjugation in $\mathbb{E}$. (That the same is true for the subgroup $\mathbb{E}^+$ of OP isometries is clear from considerations of orientation.) The third equation says that when you look at a clock through a mirror, the hands appear to move anti-clockwise. Further examples of conjugation in $\mathbb{E}$ will appear in the exercises below and in Chapter 7.

EXERCISES

2.1. Prove that any isometry of the plane maps circles to circles.

2.2. Show that all points C in the plane equidistant from two given points A and B lie on a line l, called the **perpendicular bisector** of A and B. Use this fact to give an alternative proof of the fact that any isometry of the plane maps lines to lines.

2.3. Give an example of a bijection $m: \mathbb{R}^2 \to \mathbb{R}^2$ that preserves angles but not distances. Describe in general terms the effect of m on lines, circles and triangles.

2.4. Prove that an OP isometry of $\mathbb{R}^2$ is determined by its effect on two points, and likewise for an OR isometry.

2.5. Check that translations, rotations and reflections are indeed isometries of $\mathbb{R}^2$.

2.6. A map $q: \mathbb{R}^2 \to \mathbb{R}^2$ is said to be **linear** if

$$(\mathbf{x} + \mathbf{y})q = \mathbf{x}q + \mathbf{y}q, \quad (a\mathbf{x})q = a(\mathbf{x}q),$$

for all $\mathbf{x}, \mathbf{y} \in \mathbb{R}^2$ and all $a \in \mathbb{R}$. Write down some isometries that are linear and some that are not.

2.7. Describe the effect of rotation $s(\phi)$ through ϕ about the origin on a point $P = (\rho, \theta)$ in polar coordinates.

2.8. How many elements are there in $\mathrm{Sym}(\Delta)$ when Δ is a triangle that is

(a) equilateral,   (b) isosceles,   (c) scalene?

2.9. Describe the elements of the symmetry group $\mathrm{Sym}(l)$ of a line $l \in \mathbb{R}^2$ and compare your answer with the results of Section 1.3.

2.10. Letting $\mathbb{E}^+$, $\mathbb{E}^-$ denote the sets of OP, OR isometries of $\mathbb{R}^2$ respectively, convince yourself that

$$\mathbb{E}^+ \cup \mathbb{E}^- = \mathbb{E}, \quad \mathbb{E}_+ \cap \mathbb{E}^- = \emptyset.$$

Show that for any reflection r in $\mathbb{E}$,

$$\mathbb{E}^+ r = \mathbb{E}^- = r\mathbb{E}^+.$$

2.11. Put into normal form the isometry u of $\mathbb{R}^2$ that maps the points $(0,0)$, $(1,0)$, $(0,1)$ to $(2,-1)$, $(1,-1)$, $(2,0)$ respectively.

2.12. Use the normal form theorem to give a concise description of the elements of $\mathrm{Sym}(O)$, where O is any point of $\mathbb{R}^2$. Describe explicitly the multiplication in this group.

2.13. Let s, t be non-trivial elements of $\mathbb{S}_0, \mathbb{T}$ respectively. Prove that the composite st fixes a point. Deduce that the OP subgroup $\mathbb{E}^+$ of $\mathbb{E}$ consists of translations and rotations only.

2.14. Show that the square of any OR isometry is a translation.

2.15. Given $u = r^\varepsilon st$ in normal form, put $u^{-1} = t^{-1}s^{-1}r^\varepsilon$ into normal form.

2.16. Verify that the translation subgroup $\mathbb{T}$ is closed under conjugation in $\mathbb{E}$.

2.17. Describe the effect of conjugating a rotation $s = s(\alpha)$ by a translation $t = t(\mathbf{a})$.

2.18. Let $r(l)$ denote reflection in a line l and let $u \in \mathbb{E}$ be arbitrary. Using the result of Exercise 2.2 above or otherwise, prove that $r(l)^u = r(lu)$, where lu is the image of l under u.

3

Some Basic Group Theory

Examples are the life-blood of mathematics. Among the specific examples of groups encountered so far are those in the following list.

First, the group of isometries of a set X with the discrete metric, written $\text{Sym}(X)$ and called the symmetric group on X. When X is a finite set, say $|X| = n \in \mathbb{N}$, we call $\text{Sym}(X)$ the symmetric group of degree n and denote it by S_n. S_n has $n!$ elements.

Second, there is the group $G = \text{Isom}(\mathbb{R})$ studied in Section 1.3, along with its translation subgroup G^+ and the discrete subgroup $\text{Sym}(\mathbb{Z})$.

Third, we have the Euclidean group $\mathbb{E} = \text{Isom}(\mathbb{R}^2)$ studied in the previous chapter, along with the subgroups $\mathbb{T}$ of translations and $\mathbb{S}_0$ of rotations with centre O. These last two groups are parametrised by vectors $\mathbf{a} \in \mathbb{R}^2$ and angles $\alpha \in \mathbb{R}\,\text{mod}\,2\pi$ respectively. Under the binary operation of addition inherited from $\mathbb{R}$, both $\mathbb{R}^2$ and $\mathbb{R}\,\text{mod}\,2\pi$ themselves are groups. They provide examples of two important group-theoretical constructions: direct products and factor groups respectively.

These three constructions, subgroups, factor groups and direct products, can all be used to extend our stock of examples and will be described in general terms below. But first we formalise the definition of group (Definition 1.3) and what it means to say that two groups are the same.

3.1 Groups

Definition 3.1

A **group** is a set G equipped with a binary operation, that is, a map

$$G \times G \to G, \quad (a, b) \mapsto ab,$$

such that

G1) the associative law holds:

$$(ab)c = a(bc) \quad \forall a, b, c, \in G,$$

G2) there is an identity element 1,

$$a1 = a = 1a \quad \forall a \in G,$$

G3) every element a of G has an inverse a^{-1},

$$aa^{-1} = 1 = a^{-1}a.$$

The **order** of a group G is simply its cardinality $|G|$ as a set.

Remark 3.1

The axiom G1 asserts that the product of any three elements in a group is independent of the bracketing. This can be used as the base of an induction to show that the same is true for any product of n elements, for any integer $n \geq 3$.

Remark 3.2

The axioms G2 and G3 contain some redundancy. It is sufficient to require the existence of a right identity and a right inverse:

$$a1 = a, \quad aa^{-1} = 1 \quad \forall a \in G. \tag{3.1}$$

It can be deduced from this that this identity and these inverses are two-sided, and also that they are unique.

Remark 3.3

The binary operation is denoted by concatenation and referred to as multiplication merely as a matter of convenience. It could as easily have been called addition, when we would write $a + b$ for the sum of a and b, 0 for the identity element, and $-a$ for the inverse of a. This additive notation is traditional in many cases, of which examples follow, and is standard practice in the theory of **abelian groups**, that is, groups satisfying the following extra axiom:

G4) the commutative law holds:

$$ab = ba \quad \forall a, b \in G.$$

Time for some more examples.

Each of the familiar number systems $\mathbb{Z}$, $\mathbb{Q}$, $\mathbb{R}$, $\mathbb{C}$ forms a group under addition, and the same is true by definition for any field F. Likewise the non-zero elements F^* of F form a group under multiplication, so that $\mathbb{Q}^*$, $\mathbb{R}^*$ and $\mathbb{C}^*$ are all multiplicative groups ($\mathbb{Z}^*$ is not). Again by definition, any vector space V over any field F is a group under addition, and $\mathbb{R}^2$ is an example of this kind.

The last example gives us pause. Surely any sensible list should be free of duplicates, and we already have two groups that look very much alike. The groups $\mathbb{R}^2$ and $\mathbb{C}$ under addition are "the same" in the following precise sense.

Definition 3.2

An **isomorphism** between two groups is a bijective map from one to the other that preserves the binary operation. In symbols, if G and H are groups, an isomorphism between G and H is a bijection $\phi: G \to H$ such that

$$(ab)\phi = a\phi b\phi \quad \forall a, b \in G. \tag{3.2}$$

When such a ϕ exists we write $G \cong H$ and say that G is **isomorphic** to H.

In the case of $\mathbb{R}^2$ and $\mathbb{C}$, we have the map

$$\phi: \mathbb{R}^2 \to \mathbb{C}, \quad (x, y) \mapsto x + iy,$$

mentioned in the previous chapter. This is a bijection satisfying the additive version of (3.2). Also in the previous chapter, we encountered the bijections

$$t: \mathbb{R}^2 \to \mathbb{T}, \quad \mathbf{a} \mapsto t(\mathbf{a}),$$
$$s: \mathbb{R} \bmod 2\pi \to \mathbb{S}_0, \quad \alpha \mapsto s(0, \alpha),$$

which, although written perversely on the left, are isomorphisms by the formulae (2.7) and (2.8) respectively.

An isomorphism from a group G to itself is called an **automorphism** of G. It is easy to show that the set $\mathrm{Aut}(G)$ of automorphisms of G forms a group under composition of maps. An example is conjugation by a particular element $x \in G$,

$$\gamma_x \colon G \to G, \quad g \mapsto x^{-1}gx.$$

Such automorphisms are called **inner automorphisms**, and the set of them is denoted by $\mathrm{Inn}(G)$. Simple properties of conjugation ensure that $\mathrm{Inn}(G)$ is also a group under composition of maps. Indeed the map

$$\gamma \colon G \to \mathrm{Aut}(G), \quad x \mapsto \gamma_x, \tag{3.3}$$

enjoys the property (3.2), although it is not a bijection in general; such maps are called **homomorphisms**.

3.2 Subgroups

Definition 3.3

A **subgroup** of a group G is a subset H of G that forms a group under the same binary operation as G, and then we write $H \leq G$. In symbols, $H \leq G$ if

$$1 \in H, \quad a \in H \Rightarrow a^{-1} \in H, \quad a, b \in H \Rightarrow ab \in H. \tag{3.4}$$

For example, each of the (additive) groups in the chain

$$\mathbb{Z} \leq \mathbb{Q} \leq \mathbb{R} \leq \mathbb{C}$$

is a subgroup of the next, and the examples

$$G^+, \mathrm{Sym}(\mathbb{Z}) \leq G = \mathrm{Isom}(\mathbb{R}), \quad \mathbb{T}, \mathbb{S}_0, \mathbb{E}^+ \leq \mathbb{E}$$

have already been mentioned. To get more general examples, let $\phi \colon G \to H$ be any homomorphism of groups, that is, ϕ is a map satisfying (3.2). Then the **image** and **kernel** of ϕ, defined by

$$\mathrm{Im}\,\phi = \{h \in H \mid \exists g \in G : g\phi = h\} \quad \text{and} \quad \mathrm{Ker}\,\phi = \{g \in G \mid g\phi = 1\},$$

are subgroups of H and G respectively. For the homomorphism γ in (3.3), we get

$$\mathrm{Im}\,\gamma = \mathrm{Inn}(G), \quad \mathrm{Ker}\,\gamma = \{z \in G \mid zg = gz \; \forall g \in G\}.$$

The first of these equations is just the definition of $\mathrm{Inn}(G)$, and the second holds because

$$z \in \mathrm{Ker}\,\gamma \Leftrightarrow \gamma_z = 1 \in \mathrm{Aut}(G)$$
$$\Leftrightarrow g = g\gamma_z = z^{-1}gz \quad \forall g \in G$$
$$\Leftrightarrow zg = gz \; \forall g \in G.$$

So Ker γ consists of those elements of G that commute with every element of G. This subgroup plays an important role in the general theory of abstract groups, where it is called the **centre** of G and denoted by $Z(G)$.

Another important general method for obtaining examples of subgroups depends on the elementary fact that the intersection of two, or indeed any number, of subgroups of a given group is again a subgroup.

Definition 3.4

Given any subset X of any group G, we denote by $\langle X \rangle$ the intersection of all the subgroups of G that contain X and call this the subgroup of G **generated** by X.

Given $X \leq G$, $\langle X \rangle$ is thus the smallest subgroup of G containing X, and it is not hard to show (exercise) that $\langle X \rangle$ consists of just those elements of G that can be expressed as a (finite) product of members of X and their inverses. Such elements are called **words** in $X^{\pm}$.

The extreme cases when $X = \emptyset$ and $X = G$ yield the trivial subgroup $\{1\}$ and the improper subgroup G respectively. To get a more interesting example, take X to be the set C of commutators in G, that is, those elements c of G that can be written in the form

$$c = [a, b] = a^{-1}b^{-1}ab, \quad a, b \in G.$$

The subgroup $\langle C \rangle$ is called the **derived group**, or commutator subgroup, of G and is denoted by G'.

As a final example of subgroups of this kind, take the case when the subset X of G is a singleton, say $X = \{x\}$. Then we write $\langle X \rangle = \langle x \rangle$ and say that this group is generated by the element x. Such groups, that can be generated by one element, are called **cyclic groups**. Cyclic groups have a particularly simple structure, as follows.

Fix a group G and an element $x \in G$, and denote the cyclic subgroup $\langle x \rangle$ by H. Consider the map

$$\varepsilon: \mathbb{Z} \to G, \quad k \mapsto x^k.$$

It follows from the first of the elementary rules of exponentiation,

$$x^k x^l = x^{k+l}, \quad (x^k)^l = x^{kl}, \quad k, l \in \mathbb{Z}, \tag{3.5}$$

that ε is a homomorphism. Since H consists of words in x and x^{-1}, it follows by elementary cancellation ($xx^{-1} = 1 = x^{-1}x$) that Im $\varepsilon = H$. Whether ε is injective or not is the dichotomy here, and we consider two cases.

Case 1: ε is injective. Then ε defines a bijection between its domain $\mathbb{Z}$ and its image H, and so these groups are isomorphic. In this case H is just a multiplicative version of the additive group $\mathbb{Z}$ of integers. We refer to it as the **infinite cyclic group** and denote it by Z.

Before proceeding to Case 2, note that, for each non-negative integer n, the set

$$n\mathbb{Z} = \{nk \mid k \in \mathbb{Z}\}$$

of integer multiples of n forms a subgroup of $\mathbb{Z}$ (under addition), where the cases $n = 0$ and 1 give the trivial and improper subgroup respectively. Using some elementary number theory (Euclid's division theorem) it can be verified that every subgroup of $\mathbb{Z}$ is of this form. It follows from the isomorphism that every subgroup of the infinite cyclic group $Z = \langle x \rangle$ is of the form $Z = \langle x^n \rangle$ for some integer $n \geq 0$.

Case 2: ε is not injective. This means that there are distinct $k, l \in \mathbb{Z}$ with the same image under ε: $k\varepsilon = l\varepsilon$, $k \neq l$. It follows from the homomorphism property that the kernel of ε contains the non-zero integer $k - l$. Hence, Ker ε is a non-trivial subgroup of $\mathbb{Z}$, so that Ker $\varepsilon = n\mathbb{Z}$, some $n > 0$, by the above remark. It follows that $n \in$ Ker ε, that is, $x^n = 1$. Now express any $k \in \mathbb{Z}$ in the form

$$k = nq + r, \quad 0 \leq r \leq n - 1,$$

and compute, using the rules (3.5),

$$k\varepsilon = (nq + r)\varepsilon = x^{nq+r} = (x^n)^q x^r = x^r.$$

We have shown that ε defines a bijection between the residue classes

$$r + n\mathbb{Z} = \{r + nk \mid k \in \mathbb{Z}\}, \quad 0 \leq r \leq n - 1, \tag{3.6}$$

and the elements x^r in the group $H = \langle x \rangle$. So in this case, $\langle x \rangle$ is just a multiplicative version of the group of integers modulo n under addition. We refer to it as the **cyclic group of order** n and denote it by Z_n.

As in the infinite case, the subgroups of Z_n are all cyclic: for each positive divisor d of n, $\langle x^d \rangle$ forms a subgroup of order n/d, and these comprise a complete list.

Finally, a piece of nomenclature. Given an element x in a group G, the **order** $|x|$ of x is the number of its distinct powers, that is, the order of the group $\langle x \rangle$, or the cardinality of $\langle x \rangle$ as a set. Thus, either $|x| = \infty$ or $|x| = n \in \mathbb{N}$ according to case. In the latter case, n is the smallest positive integer for which $x^n = 1$, and the above analysis shows that

$$\text{for } m \in \mathbb{Z}, \ x^m = 1 \Leftrightarrow m \text{ is a multiple of } |x|.$$

3.3 Factor Groups

The prime example of a factor group is the group of integers modulo n under addition. Its elements are not integers but sets of integers, namely, the residue classes in (3.6). The crucial property of residue classes is that they form a **partition** of $\mathbb{Z}$: every integer belongs to exactly one residue class.

There is a similar situation in geometry, as follows. Let l be a line through the origin $\mathbf{0}$ in $\mathbb{R}^2$. Then the image of l under a translation $t = t(\mathbf{a})$,

$$lt = l + \mathbf{a} = \{\mathbf{x} + \mathbf{a} \mid \mathbf{x} \in l\},$$

is again a line. In fact, it is the line parallel to l through the point $\mathbf{a} \in \mathbb{R}^2$. Being parallel, two such lines are either equal of disjoint:

$$l + \mathbf{a} = l + \mathbf{b} \Leftrightarrow \mathbf{a} - \mathbf{b} \in l, \text{ and } (l + \mathbf{a}) \cap (l + \mathbf{b}) = \emptyset \text{ otherwise.}$$

So these lines form a partition of $\mathbb{R}^2$: every point in the plane belongs to exactly one of them. In fact, $\mathbf{a} \in \mathbf{a} + l$ (since $\mathbf{0} \in l$).

Both of these examples are special cases of a general situation in group theory, the common element being the subgroup property: $n\mathbb{Z}$ is a subgroup of $\mathbb{Z}$ and l is a subgroup of $\mathbb{R}^2$, under addition of integers and vectors respectively. To describe this general situation, let us fix a group G and a subgroup $H \leq G$ and revert to multiplicative notation. Proceeding in multiplicative analogy with the above, we define

$$Hx = \{hx \mid h \in H\}, \quad x \in G,$$

and call this subset a **right coset** of H in G. To show that the right cosets of H partition G, we proceed as follows.

First, if $x \in H$, note that multiplication on the right by x simply permutes the elements of H. This is the essential content of Cayley's theorem (Exercise 3.16). So Hx and H are equal as sets in this case. Next, if $x \notin H$, then $x = 1x \in Hx$, and Hx and H are not equal as sets. Hence,

$$Hx = H \Leftrightarrow x \in H,$$

and it follows easily from this that, for $x, y \in G$,

$$Hx = Hy \Leftrightarrow xy^{-1} \in H. \tag{3.7}$$

Now assume that Hx and Hy are not disjoint, say $z \in Hx \cap Hy$. Then we have $zx^{-1}, zy^{-1} \in H$, and so

$$Hx = Hz = Hy.$$

We have shown that two cosets are either equal or disjoint, and so the cosets partition G as claimed. It is just the partition obtained from the equivalence relation $x \sim y$ if $xy^{-1} \in H$. We can now refer to Hx more precisely as the right coset of H containing x.

The number of right cosets of H in G is called the **index** of H in G and written $|G : H|$. Since the map $H \to Hx$, $h \mapsto hx$, is clearly a bijection, all right cosets of H have the same number of elements, and we have proved the following classical result.

Theorem 3.1 (Lagrange)

If H is a subgroup of a group G, then

$$|G| = |G : H|\,|H|. \tag{3.8}$$

□

Now denote the set of (distinct) right cosets of H in G by $\mathcal{C}$ and let T be a subset of G compiled by choosing exactly one element from each such coset; T is called a **right transversal** for H in G. (Note that when neither $|H|$ nor $|G : H|$ is finite, appeal must be made here to the Axiom of Choice.) Then $|T| = |G : H|$ and every $g \in G$ admits a unique decomposition

$$g = ht, \quad h \in H, \quad t \in T.$$

Thus, sitting behind the factorisation (3.8) of $|G|$ as a product of two smaller numbers (assuming that $|G|$ is finite and H is proper and non-trivial), we have a decomposition $G = HT$, one of which is a group. The study of G can thus be broken down into the study of two smaller groups under either of the following two conditions:

C1) the coset representatives can be chosen in such a way that T is a group,

C2) the natural multiplication $(Hx)(Hy) = H(xy)$ makes $\mathcal{C}$ into a group.

When C1 holds, T is called a **complement** of H in G, and the decomposition $G = HT$ is called a **factorisation** of G. While this condition has its uses, it turns out that the condition C2 is vastly more fruitful, and we take a close look at it now.

Note first of all that the whole business with right cosets can be carried out equally well with left cosets

$$xH = \{xh \mid h \in H\}, \quad x \in G,$$

instead. (Note that the number of left cosets is equal to the number of right cosets, so the index $|G : H|$ is well defined.) Now observe that the problem with the equation in C2 is one of definition: the coset $H(xy)$ may depend on the choice of representatives x, y from the cosets Hx, Hy respectively. In a nutshell, for a different choice, say $x' = h_1 x \in Hx$, $y' = h_2 y \in Hy$, where $h_1, h_2 \in H$, we may get a different coset $H(x'y')$. For C2 to constitute a good definition, the requirement is therefore that

$$H(x'y') = H(xy)$$

for all such choices of x' and y'. According to (3.7), this means that the element $x'y'(xy)^{-1} = x'y'y^{-1}x^{-1} = h_1 x h_2 x^{-1}$ must belong to H for all $h_1, h_2 \in H$ and all $x \in G$, and this is so if and only if $x h_2 x^{-1} \in H \ \forall h_2 \in H, \forall x \in G$, that is, H is closed under conjugation by elements of G. Then $x^{-1}Hx \subseteq H$ and also

$$xHx^{-1} \subseteq H \Rightarrow H = x^{-1}xHx^{-1}x \subseteq x^{-1}Hx,$$

and so we require that

$$H^x = H, \text{ or equivalently } Hx = xH, \quad \forall x \in G.$$

It is a sad fact that this condition on the subgroup H does not hold in general, but when it does, it is a matter of routine to check the group axioms for $\mathcal{C}$.

Definition 3.5

A subgroup H of a group G is said to be **normal** if $Hx = xH$ for all $x \in G$, and in this case we write $H \triangleleft G$. Given $H \triangleleft G$ the subgroup formed by the cosets of H in G under the operation $(Hx)(Hy) = H(xy)$, $x, y \in G$, is called the **factor group** of G by H and written G/H.

We give three examples, two specific and one of general importance.

Example 3.1

Let G be additive group $\mathbb{Z}$ and H the subgroup $n\mathbb{Z}$, $n \in \mathbb{N}$, necessarily normal as $\mathbb{Z}$ is abelian. Then the cosets are just the residue classes modulo n, and they constitute the factor group $\mathbb{Z}/n\mathbb{Z}$ of integers modulo n. Since this group is cyclic (with generator $1 + \mathbb{Z}/n\mathbb{Z}$), we have $\mathbb{Z}/n\mathbb{Z} \cong Z_n$. This is the result of applying the fundamental isomorphism theorem of group theory (see Exercise 3.8) to the map ε studied in Section 3.2.

Applying the same theorem to the map $\mathbb{R} \to \mathbb{E}$, $\alpha \mapsto s(O, \alpha)$, we get our next example.

Example 3.2

Taking $G = \mathbb{R}$ and $H = 2\pi\mathbb{Z}$, so that $H \lhd G$ for the same reason as in Example 3.1, we get the factor group $\mathbb{R}/2\pi\mathbb{Z}$, previously denoted by $\mathbb{R} \bmod 2\pi$. As in the previous example, elements of the subgroup ($n\mathbb{Z}$ or $2\pi\mathbb{Z}$) are "regarded as" zero or trivial. Such treatment is made rigorous by the definition of the factor group; indeed, the identity element of the factor group G/H is just the trivial coset H.

Example 3.3

For our last example, we let G be arbitrary and take H to be the derived group G'. Since conjugation preserves commutators, $[x,y]^z = [x^z, y^z]\ \forall x, y, z \in G$, and hence their inverses and products, it follows that $G' \lhd G$. The factor group G/G' is called the **derived factor group** of G, or G abelianised and written G^{ab}, for the following reason. Given a normal subgroup H of G, the left and right cosets of H are the same, and then

$$G/H \text{ is abelian} \Leftrightarrow (xH)(yH) = (yH)(xH) \quad \forall x, y \in G$$
$$\Leftrightarrow (xy)H = (yx)H \quad \forall x, y \in G$$
$$\Leftrightarrow (x^{-1}y^{-1}xy)H = H \quad \forall x, y \in G$$
$$\Leftrightarrow [x, y] = x^{-1}y^{-1}xy \in H \quad \forall x, y \in G$$
$$\Leftrightarrow G' \leq H.$$

Thus, $G^{ab} = G/G'$ is, in this precise sense, the biggest abelian factor group of G. We shall have more to say about abelianisation in Chapter 5.

3.4 Semidirect Products

For our third and final basic construction, we take the favourable case when both conditions C1 and C2 in Section 3.3 hold for a subgroup H of a group G. Thus, we assume that H is normal and has a complement in G. To avoid confusion with translation subgroups we rename this complement K and, also for future convenience, we redenote the resulting decomposition $G = KH$, which is permissible as $H \lhd G$. In this situation, we refer to G as a **semidirect product** of H and K. The indefinite article is used here because the structure of G is not determined by H and K alone, and we investigate this phenomenon now.

Let g, g' be two elements of G, with factorisations

$$g = kh, \quad g' = k'h', \quad k, k' \in K, \quad h, h' \in H.$$

Then their product has a similar factorisation, which is calculated as follows:

$$g'g = k'h'kh = k'kh'^k h,$$

with $k'k \in K$ and $h'^k h \in H$, since H is closed under conjugation. So the structure of G depends on a third ingredient (in addition to H and K), namely, the action of K on H by conjugation. This action is neatly expressed as a homomorphism

$$\alpha: K \to \operatorname{Aut}(H), \quad k \mapsto \alpha_k,$$

where

$$\alpha_k: H \to H, \quad h \mapsto h^k,$$

is the automorphism of conjugation by k. The standard properties of conjugation guarantee that α_k is indeed an automorphism of H and that α is a homomorphism: $\alpha_k \alpha_{k'} = \alpha_{kk'} \ \forall k, k' \in K$.

We shall now invert this process to obtain the desired construction. The input is a triple (H, K, α), where H and K are groups and $\alpha: K \to \operatorname{Aut}(H)$, $k \mapsto \alpha_k$, is a homomorphism. Then a binary operation on the Cartesian product $K \times H$ is defined as follows: for $(k, h), (k', h') \in K \times H$, set

$$(k', h')(k, h) = (k'k, h'\alpha_k h), \tag{3.9}$$

where $h'\alpha_k$ is the image of $h' \in H$ under the automorphism α_k. It is a matter of routine to check the group axioms, and we denote the resulting group by $K \times_\alpha H$. Regarding H and K as embedded in this group via the maps $h \mapsto (1, h)$ and $k \mapsto (k, 1)$ respectively, it is again routine to check that H is a normal subgroup with complement K, with the action of K on H by conjugation being given by α. This completes the construction of the semidirect product $K \times_\alpha H$ of H and K with respect to α. Time for some examples.

Example 3.4

Any abelian group A admits the automorphism ι of inversion (Exercise 3.10), $a\iota = a^{-1} \ \forall a \in A$. Since $\iota^2 = 1$, the map $\alpha: Z_2 \to \operatorname{Aut}(A)$, $1 \mapsto 1$, $x \mapsto \iota$, is a homomorphism. Abusing notation to the extent of writing ι in place of α, we describe the semidirect product $Z_2 \times_\iota A$ for various A.

(a) When $A = \mathbb{R}$, the group of real numbers under addition, we recognise $Z_2 \times_\iota A$ as the group $G = \operatorname{Isom}(\mathbb{R}, d)$ studied in Section 1.3:

$$\operatorname{Isom}(\mathbb{R}, d) \cong Z_2 \times_\iota \mathbb{R}.$$

These groups are isomorphic because, in the former, reflection r in 0 has order 2 and acts by inversion on the translation subgroup $T \cong \mathbb{R}$, and this is precisely the recipe used to construct the latter.

(b) When $A = \mathbb{Z}$, the group of integers under addition, the same reasoning enables us to recognise $Z_2 \times_\iota A$ as the infinite dihedral group (see formula (1.16) in Chapter 1):

$$D_\infty \cong Z_2 \times_\iota \mathbb{Z}.$$

(c) When $A = \mathbb{R}/2\pi\mathbb{Z}$, it follows from the corollary to Theorem 2.4 (the last equation in (2.14)) that $Z_2 \times_\iota A$ is the subgroup $\mathbb{S}_0$ of the Euclidean group,

$$\mathbb{S}_0 \cong Z_2 \times_\iota \mathbb{R}/2\pi\mathbb{Z}.$$

(d) For a new example, take $A = Z_n = \langle y \rangle$ and let $Z_2 = \{1, x\}$, with x again acting by inversion. This gives the relations

$$y^n = x^2 = 1, \quad x^{-1}yx = y^{-1}$$

defining the **dihedral group of order** $2n$ on the generators x and y. We write

$$D_{2n} = \langle x, y \mid y^n = x^2 = 1, x^{-1}yx = y^{-1} \rangle, \tag{3.10}$$

so that

$$D_{2n} \cong Z_2 \times_\iota Z_n.$$

The geometrical significance of this group is the subject of Exercise 3.24, and this gives the reason for its name.

Example 3.5

Consider the subgroups

$$\mathbb{T} \cong \mathbb{R}^2, \quad \mathbb{S}_0 \cong \mathbb{R}/2\pi\mathbb{Z}$$

of the Euclidean group $\mathbb{E}$. It follows from the Normal Form Theorem that

$$\mathbb{S}_0\mathbb{T} = \mathbb{E}^+, \quad \mathbb{S}_0 \cap \mathbb{T} = \{1\}.$$

Also, by the corollary to Theorem 2.4 (the second equation in (2.14)), we know that $\mathbb{T}$ is a normal subgroup of $\mathbb{E}^+$. This leads us to recognise $\mathbb{E}^+$ as a semidirect product $\mathbb{S}_0 \times_\sigma \mathbb{T}$, where, for $s \in \mathbb{S}_0$, $s\sigma$ is the automorphism of $\mathbb{T}$ sending $t(\mathbf{a})$ to $t(\mathbf{a}s)$.

Now let $r \in \mathbb{E}$ denote reflection in any axis through 0 and denote the group $\langle r \rangle = \{1, r\}$ by Z_2. We have

$$Z_2\mathbb{E}^+ = \mathbb{E}, \quad Z_2 \cap \mathbb{E}^+ = \{1\},$$

and the normality of $\mathbb{E}^+$ in $\mathbb{E}$ follows from considerations of orientation. The action of r on $\mathbb{T}$ and $\mathbb{S}_0$, and hence on $\mathbb{E}^+$, is given by the first and third equations in (2.14), and we see that $\mathbb{E}$ is a semidirect product $Z_2 \times_\rho \mathbb{E}^+$, where $r\rho$ is the automorphism of $\mathbb{E}^+$ sending $s(\beta)t(\mathbf{a})$ to $s(-\beta)t(\mathbf{a}r)$.

So we see that $\mathbb{E}$ is a sort of twofold semidirect product, constructed using the basic isometries of translation, rotation and reflection in two steps:

(a) extend $\mathbb{T}$ by $\mathbb{S}_0$ to get $\mathbb{E}^+$,

(b) extend $\mathbb{E}^+$ by $\mathbb{Z}_2 = \langle r \rangle$ to get $\mathbb{E}$.

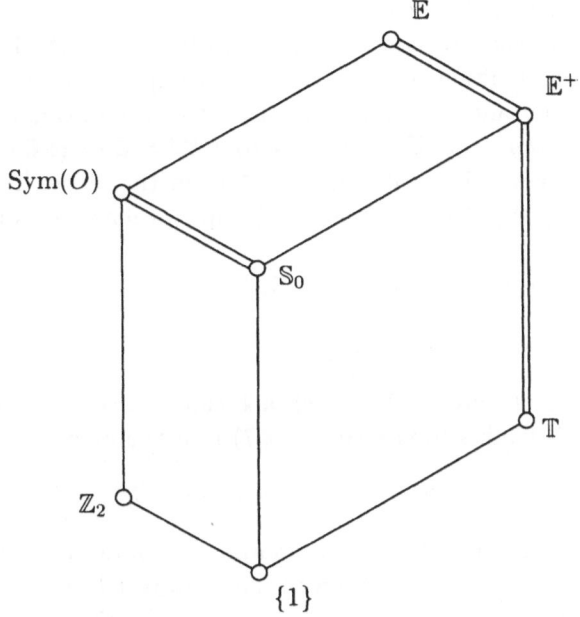

Figure 3.1   Hasse diagram for $\mathbb{E}$

Group theorists like to represent this kind of situation by what they call a
Hasse diagram. Referring to Fig. 3.1, the diagram is meant to represent a
portion of the subgroup lattice of a group, $\mathbb{E}$ in this case. The small circles
represent subgroups, two being joined by a line segment when the lower is
contained in the higher. The whole group $\mathbb{E}$ is thus at the top, and the trivial
group $\{1\}$ at the bottom. A double line segment indicates normality of the lower
subgroup in the higher and, in a parallelogram, the highest subgroup is the
product, and the lowest the intersection, of the other two. This is illustrated by
the subgroup $H = \langle r \rangle \mathbb{S}_0$, which is just the group $\mathrm{Sym}(0) \cong \mathbb{Z}_2 \times_\iota \mathbb{S}_0$ mentioned
in the previous example.

Example 3.6

Returning briefly to the general case $G = K \times_\alpha H$, an important special case is
when the homomorphism $\alpha \colon K \to \mathrm{Aut}(H)$ is trivial: $k\alpha = 1$, the identity map,

for all $k \in K$. Equation (3.9) then reads

$$(k', h')(k, h) = (k'k, h'h),$$

that is, multiplication is componentwise. The resulting group is called the **direct product** of K and H, written $K \times H$.

An example already to hand is the group $\mathbb{R}^2 = \mathbb{R} \times \mathbb{R}$. Since the binary operation in $\mathbb{R}$ is addition, that in $\mathbb{R} \times \mathbb{R}$ is componentwise addition, which is the usual rule for addition of vectors. To get another example, consider the subgroup $\mathbb{Z}$ of integers in $\mathbb{R}$. Then the subgroup $\mathbb{Z}^2 = \mathbb{Z} \times \mathbb{Z}$ of $\mathbb{R} \times \mathbb{R}$ corresponds to the set of integer points, (m, n), $m, n \in \mathbb{Z}$, in the plane. To get hold of a multiplicative version of this important group, we make use of isometries as follows.

Consider the translations $a = t(1, 0)$, $b = t(0, 1) \in \mathbb{E}$. Since translations commute,

$$ab = t(1, 1) = ba, \tag{3.11}$$

any element of the subgroup $D = \langle a, b \rangle$ of $\mathbb{E}$ can be written in the form $a^m b^n$, $m, n \in \mathbb{Z}$. Then it follows from formula (2.7) that the map

$$t : \mathbb{Z} \times \mathbb{Z} \to D, \quad (m, n) \mapsto a^m b^n,$$

written perversely on the left, is an isomorphism. Moreover, the form $a^m b^n$ is unique, and so every relation between the generators a, b of D is a consequence of (3.11). We thus identify D as the abstract group with presentation

$$Z^2 = \langle a, b \mid [a, b] = 1 \rangle. \tag{3.12}$$

This group will play a fundamental role in later chapters; it is called the **free abelian group of rank** 2.

Although defined for arbitrary groups, the direct product construction is particularly important in the case of abelian groups. Since cyclic groups are abelian, and the direct product of two abelian groups is abelian, an easy induction shows that the group

$$A = Z_{d_1} \times Z_{d_2} \times \cdots \times Z_{d_n} \tag{3.13}$$

is abelian, where Z_{d_i} is the cyclic group of order $d_i \in \mathbb{N}$, $1 \leq i \leq n$. The rather surprising fact is that *every* finite abelian group is of this form. This assertion is one half of the classification theorem for finite abelian groups. The other half, which is equally non-obvious, consists of saying exactly when two decompositions of the form (3.13) represent isomorphic groups. Both halves will be proved in Chapter 5 below.

EXERCISES

3.1. Prove the generalised associative law for groups: for elements x_1, x_2, ..., x_n of a group, their product in this order is independent of the order in which the $n - 1$ multiplications are carried out.

3.2. Given a set G with an associative binary operation for which there is a right identity e and a right inverse x' for every element x:

$$xe = x, \quad xx' = e,$$

prove that the identity and inverses are two-sided and unique. Deduce that G is a group.

[**Hint.** Begin by evaluating the product $x'xx'(x')'$ in two different ways.]

3.3. Prove that the **cancellation laws**

$$xa = ya \Rightarrow x = y, \quad ax = ay \Rightarrow x = y$$

hold in any group.

3.4. Say why the non-zero integers do not form a group under multiplication.

3.5. Use a suitable elementary function to prove that the group $\mathbb{R}$ of reals under addition is isomorphic to the group $\mathbb{R}_+$ of positive reals under multiplication.

3.6. Given isomorphisms $\theta: G \to H$, $\phi: H \to K$ of groups, show that $\theta\phi: G \to K, \theta^{-1}: H \to G$ are also isomorphisms. Deduce that $\text{Aut}(G)$ is a group under composition of maps.

3.7. Prove that the map $\gamma_x: G \to G$ of conjugation by a fixed $x \in G$ is an automorphism. Prove that the map $\gamma: G \to \text{Aut}(G)$, $x \mapsto \gamma_x$, is a homomorphism.

3.8. (Fundamental isomorphism theorem) Given a homomorphism $\phi: G \to H$ of groups, show that its kernel Ker ϕ and image Im ϕ are subgroups of G and H respectively. Show further that the subgroup $K = \text{Ker } \phi$ is a normal subgroup of G, and that the map $\phi': G/K \to \text{Im } \phi$, $Kx \mapsto x\phi$, defines an isomorphism of groups.

3.9. With $\gamma: G \to \text{Aut}(G)$ as in Exercise 3.7, show that Ker γ is equal to the centre $Z(G)$ of G. Show further that the image $\text{Inn}(G)$ of γ is a normal subgroup of $\text{Aut}(G)$.

3.10. Given a group G, prove that the map $\iota\colon G \to G$, $x \mapsto x^{-1}$, is an automorphism if and only if G is abelian.

3.11. Prove that any intersection of subgroups of a group is again a subgroup.

3.12. Given a subset X of a group G, let $W(X)$ denote the set of finite products of elements of X and their inverses. Prove that $W(X) = \langle X \rangle$.

3.13. Let H be a non-zero subgroup of the group $\mathbb{Z}$ of integers under addition. Prove that $H = n\mathbb{Z}$, where n is the smallest positive element of H.

3.14. Prove that every subgroup of the cyclic group $Z_n = \langle x \mid x^n = 1 \rangle$, $n \in \mathbb{N}$, is of the form $\langle x^d \rangle$, where d is a divisor of n.

3.15. Given a subgroup H of a group G, show that the relation $x \sim y$ if and only if $xy^{-1} \in H$ is an equivalence relation on G. Show that the resulting equivalence classes are just the right cosets of H in G. Find an equivalence relation that similarly gives rise to the left cosets of H in G.

3.16. (Cayley's theorem) Let G be a group with underlying set $\widehat{G}$, and let $\mathrm{Sym}(\widehat{G})$ denote the group of all permutations of $\widehat{G}$ (bijections from $\widehat{G}$ to itself, with composition of maps as the binary operation). Show that the map $\rho_x\colon \widehat{G} \to \widehat{G}$, $g \mapsto gx$, where $x \in G$, belongs to $\mathrm{Sym}(\widehat{G})$, and that the rule $\rho\colon G \to \mathrm{Sym}(\widehat{G})$, $x \mapsto \rho_x$, defines an injective homomorphism of groups.

3.17. (The orbit-stabiliser theorem) Let A be a finite set and G a subgroup of $\mathrm{Sym}(A)$. For a fixed $a \in A$ define the G-**orbit** of a as

$$aG = \{b \in A \mid b = ax \text{ for some } x \in G\}$$

and the **stabiliser** of a in G as

$$\mathrm{Stab}_G(a) = \{x \in G \mid ax = a\}.$$

Then prove that $\mathrm{Stab}_G(a)$ is a subgroup of G of index $|aG|$.

3.18. Use Lagrange's theorem to prove that if x is an element of a finite group G, then $|x|$ is a divisor of $|G|$. Deduce that every group of prime order is cyclic.

3.19. Given $H \leq G$, prove that the number of left cosets of H in G is equal to the number of right cosets.

3.20. Let H be a subgroup of finite index in a group G, and let $K \leq G$ with $H \leq K$. Prove that $|K : H||G : K| = |G : H|$ without using Lagrange's theorem.

3.21. Prove that subgroups of index 2 are always normal.

3.22. Check that when $H \lhd G$ the multiplication on $\mathcal{C}$ defined in condition C2 does indeed satisfy the group axioms.

3.23. Consider the infinite dihedral group

$$D_\infty = \langle x, y \mid x^2 = 1, x^{-1}yx = y^{-1} \rangle$$

(see formula (1.16) and Example 3.4(b)). Describe the derived group D'_∞ and derived factor group D_∞^{ab}. The same for the finite dihedral group D_{2n}, $n \in \mathbb{N}$ (see Example 3.4(d)).

3.24. Let P_n be a regular plane n-gon, $n \in \mathbb{N}$, $n \geq 3$. Prove that $\mathrm{Sym}(P_n) \cong D_{2n}$. Find a subgroup H of this group that is not normal.

3.25. Check that the multiplication in $K \times_\alpha H$ defined by (3.9) satisfies the group axioms.

3.26. Let $H \lhd G$ with complement K. Prove that the induced action of K on H is trivial ($h^k = h$, $\forall h \in H$, $k \in K$) if and only if K is also normal in G, and that then $G \cong K \times H$.

3.27. Prove that the direct product of a finite number of cyclic groups is abelian.

3.28. Prove that the subgroup $D = \langle t(1,0), t(0,1) \rangle$ of $\mathbb{E}$ is discrete.

3.29. Prove that the subgroup $\langle y^2, x \rangle$ of the group D_8 is isomorphic to the direct product $Z_2 \times Z_2$.

3.30. Prove that $Z_2 \times Z_3 \cong Z_6$.

3.31. Describe each of the four automorphisms of Z_8 by specifying its effect on a generator. Show that the corresponding four groups of the form $Z_2 \times_\alpha Z_8$ are pairwise non-isomorphic.

4

Products of Reflections

With a sigh of relief we take a welcome break from the abstraction and formalism of group theory and return to the refreshingly picturesque world of plane geometry.

In Chapter 2 the Normal Form Theorem was used to make some progress towards understanding the structure of the Euclidean group $\mathbb{E}$. An alternative method, which perhaps yields more in the way of geometric insight, is described in this chapter. It can be used, for example, to describe the product of two non-concentric rotations, to give a complete classification of plane isometries and a characterisation of them in terms of fixed points and orientation, and to guide the first few tentative steps towards a study of the isometries of $\mathbb{R}^3$. All this simply by answering the question "what is the product of two reflections?"

4.1 The Product of Two Reflections

Let r and r' be the reflections in lines l, l' respectively. Since a reflection is determined by its axis and vice versa, it follows that $r = r'$ if and only if $l = l'$. We discard the trivial case $l = l'$, when $rr' = r^2 = 1$, and assume at once that $l \neq l'$. We distinguish two cases according as l and l' meet (say in a point O) or not. These are illustrated in Fig. 4.1, where the angle from l to l' is α in case (i) and the perpendicular distance from l to l' is a in case (ii).

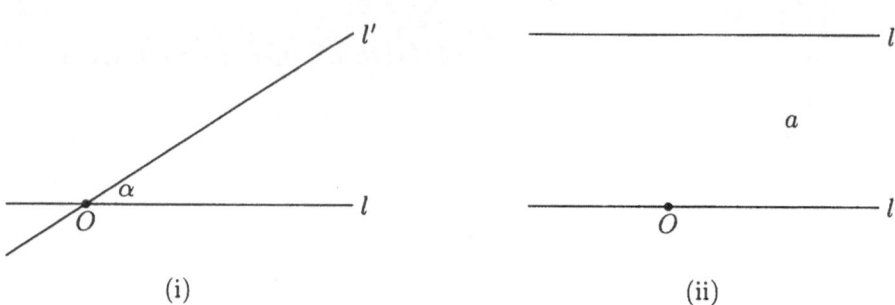

Figure 4.1   Two reflections

Case (i): $l \cap l' = \{O\}$. In polar coordinates with centre O, base line l and $\alpha > 0$, an arbitrary point $P = (\rho, \theta)$ is mapped by r to $(\rho, -\theta)$, while r' sends P to the point $P' = (\rho, \phi)$, where $\frac{\theta + \phi}{2} = \alpha$, since l' bisects the angle $P\hat{O}Pr'$. Thus, $\phi = 2\alpha - \theta$, and we compute

$$(\rho, \theta)rr' = (\rho, -\theta)r' = (\rho, 2\alpha + \theta).$$

The product rr' thus fixes the distance from O and increases the polar angle by 2α. That is, $rr' = s(O, 2\alpha)$, rotation about O through 2α.

Case (ii): $l \| l'$. In Cartesian coordinates with origin O, x-axis l and $a > 0$, an arbitrary point $P = (x, y)$ is mapped by r to $(x, -y)$, while r' sends P to the point $P' = (x, z)$, where $\frac{y+z}{2} = a$, since l' is the perpendicular bisector of P and P'. Thus, $z = 2a - y$, and we compute

$$(x, y)rr' = (x, -y)r' = (x, 2a + y).$$

The product rr' thus fixes the x-coordinate and increases the y-coordinate by $2a$. That is, $rr' = t(0, 2a)$, translation by $2a$ in the direction of the normal from l to l'.

Before stating these results as a theorem, we pause to admire the wonderful similarity of treatment in the two cases.

Theorem 4.1

Given reflections r, r' in distinct lines l, l' in $\mathbb{R}^2$, their product rr' is a rotation or a translation according as l, l' meet or not. In the former case the centre of rotation is the point of intersection and the angle twice that from l to l'. In the latter case the direction of translation is that of the normal from l to l' and the distance twice that from l to l'.   □

An immediate consequence of this extremely useful theorem is its equally useful converse: any rotation or translation of $\mathbb{R}^2$ is the product of two reflections. Specifically,

$$s(O, \alpha) = rr',$$

where the axes of r, r' may be chosen as any two lines l, l' through O such that the angle between them, measured from l to l', is $\alpha/2$. Also

$$t(\mathbf{a}) = rr',$$

where the axes of r, r' may be chosen as any two lines l, l' perpendicular to the direction of $\mathbf{a}$ such that the distance between them, measured from l to l', is half the magnitude of $\mathbf{a}$. We emphasize that in each case, the latitude in the choice of r and r' is especially valuable, as we shall see in a moment.

4.2 Three Reflections

We shall compute the product of three reflections r_1, r_2, r_3 with axes l_1, l_2, l_3. There is a veritable proliferation of cases here, according to the many possibilities for the distribution of the lines l_1, l_2, l_3 in $\mathbb{R}^2$. But the judicious application of Theorem 4.1 drastically reduces the number of these cases, and we are essentially left with only one outstanding.

Since two lines can have at most one point in common, the total number n of points of intersection of pairs of distinct lines among l_1, l_2, l_3 is at most three. Thus we distinguish four cases according as $n = 0, 1, 2$ or 3. The first two of these are depicted in Fig. 4.2 and include the degenerate cases when two or more of l_1, l_2, l_3 are the same. Let r denote reflection in the line l chosen in each case so that $r_1 r_2 = r r_3$, in accordance with the latitude in the two parts of the converse of Theorem 4.1. Then in both cases, $r_1 r_2 r_3 = r r_3 r_3 = r$ is again a reflection.

The remaining cases $n = 2, 3$ are illustrated in Fig. 4.3. Note first that the case $n = 3$ is a "symmetric" one: l_1, l_2, l_3 label the three sides of a triangle in (necessarily) cyclic order. Next observe that Fig. 4.3(ii) can be transformed into Fig. 4.3(i) by simultaneously rotating l_1 and l_2 about their point of intersection while keeping l_3 fixed. As above, this operation does not affect the product $r_1 r_2$. Similar operations convert Fig. 4.3(i) into the other two subcases, $l_1 \| l_2$ and $l_1 \| l_3$, of the case $n = 2$. It therefore remains only to calculate $r_1 r_2 r_3$ in the case $n = 2$ depicted in Fig. 4.3(i). For this we shall need the result in Theorem 2.2.

With this in mind, we annotate and expand Fig. 4.3(i) into Fig. 4.4, where O is the point of intersection of l_1 and l_2, α is the angle from l_1 to l_2, a is the distance from l_2 to l_3, and l is the line through O perpendicular to l_2. For ease

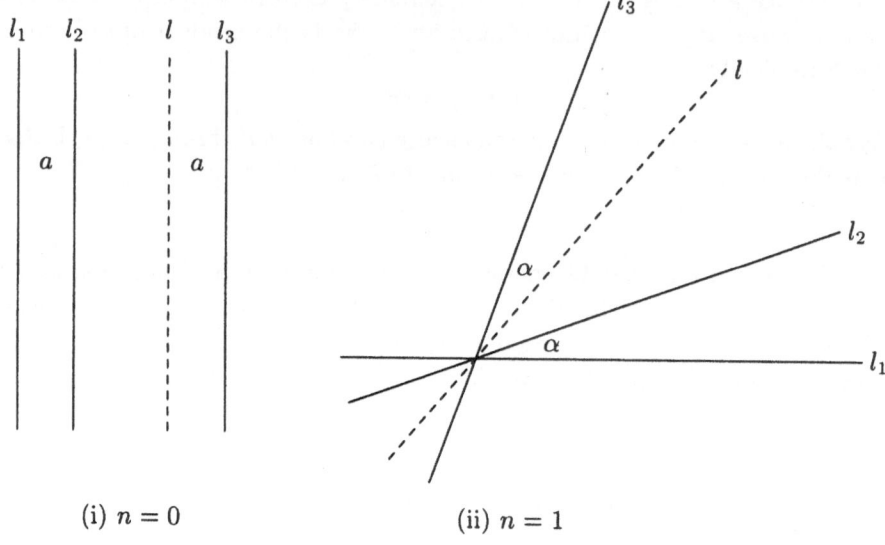

(i) $n = 0$ (ii) $n = 1$

Figure 4.2   Three reflections: $n = 0$ or 1

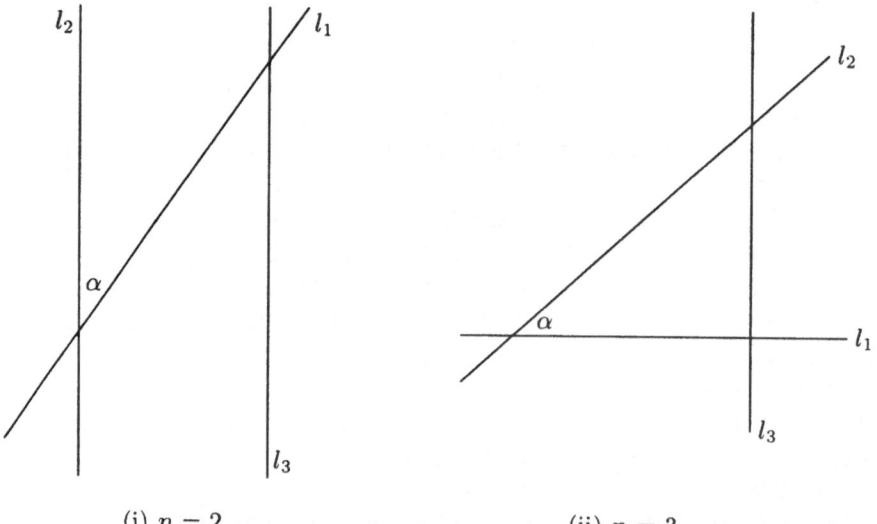

(i) $n = 2$ (ii) $n = 3$

Figure 4.3   Three reflections: $n = 2$ or 3

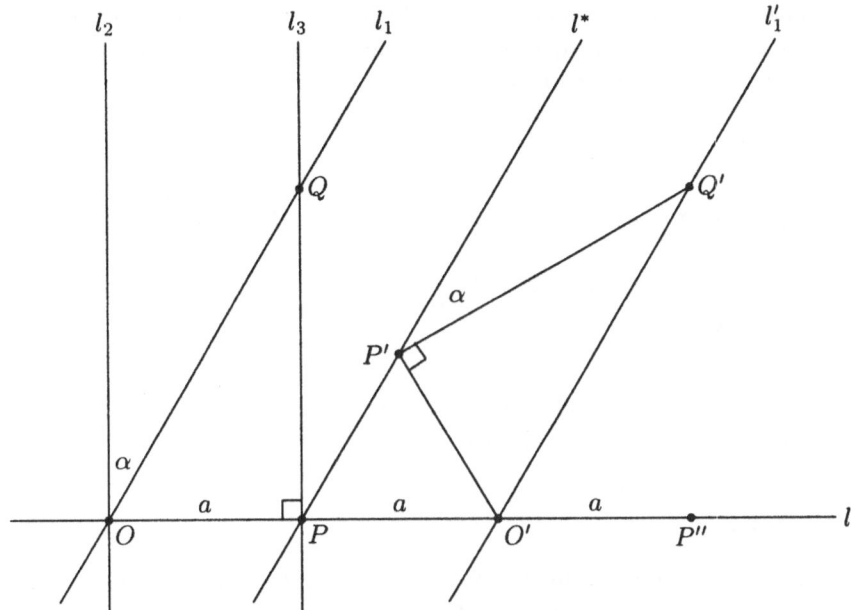

Figure 4.4   The one remaining case

of reference, we think of O as the origin of coordinates and l, l_2 as the x-, y-axes respectively. Letting P, Q be the points of intersection of l_3 with l, l_1 respectively, we distinguish two cases.

Case (i): $P \neq Q$, so that O, P, Q are not collinear since $a \neq 0$. We proceed to find the images O', P', Q' of O, P, Q under $r_1 r_2 r_3$. First, since O and Q are fixed by r_1, it follows that O' and Q' are their images under the translation $r_2 r_3$, which moves them through a distance of $2a$ to the right. Next, by Theorem 2.1, the triangles $OPQ, O'P'Q'$ are congruent (SSS). There are thus two candidates for P', either the point P'' with Cartesian coordinates $(3a, 0)$ or its image under reflection in the line l_1' through O', Q'. But since $r_1 r_2 r_3$ is OR, the second candidate is the winner.

Now let l^* be the line through P, P' and denote by q the composite isometry of translation along l^* from P to P' followed by reflection in l^*. Inspection of Fig. 4.4 reveals that $Pq = P', Qq = Q', Oq = O'$, and we deduce from Theorem 2.2 that $r_1 r_2 r_3 = q$. A similar situation arises in the second case, as follows.

Case (ii): $P = Q$, so that $l = l_1$ is perpendicular to l_2 and l_3, and $r_1 r_2 r_3$ is reflection $r = r_1$ in l followed by a translation $t = r_2 r_3$ along it. Now observe that rt and tr both send an arbitrary point (x, y) to $(x + 2a, -y)$. Thus, $r_1 r_2 r_3 = rt = tr$, and the situation is the same as in Case (i). Notice that, since $a \neq 0$, the isometry rt has no fixed points, and so cannot be a reflection. And being OR, it is not a translation or a rotation either. We have thus discovered a new type of isometry of $\mathbb{R}^2$.

Definition 4.1

Given a pair P, P' of distinct points on a line $l \in \mathbb{R}^2$, the isometry

$$q(P, P') := r(l)t(\overrightarrow{PP'})$$

is called a **glide reflection**.

As already observed, such isometries have no fixed points and are OR, and we shall see that these two properties characterise glide reflections.

Theorem 4.2

The product of three reflections in $\mathbb{E}$ is either a reflection or a glide reflection according as the number of points of intersection of distinct axes is less than or greater than $3/2$. $\square$

4.3 Four or More

First consider the product u of four reflections. By Theorem 4.1, u is the product of two translations, two rotations or one of each. The first possibility has already been treated: according to Chapter 2, the result is again a translation. Next let $u = st$, where s is a rotation, t is a translation, and both may be assumed to be non-trivial. Exericse 2.13 asks for a proof that u is again a rotation; here is a slick and constructive solution.

Let $s = s(O, \alpha)$, $t = t(\mathbf{a})$, and construct three lines in $\mathbb{R}^2$ as follows:

- l is the line through O perpendicular to the direction of $\mathbf{a}$,
- l' is the line through O such that the angle from l' to l is $\alpha/2$,
- l'' is the perpendicular bisector of O and Ot.

Then by Theorem 4.1,

$$s = r(l')r(l), \quad t = r(l)r(l''),$$

whence

$$u = st = r(l')r(l'')$$

as $r(l)^2 = 1$. Since l' and l'' are not parallel, this is a rotation, by Theorem 4.1 again. The reader is invited to supply an appropriate diagram.

Moving on to the next case, $u = ts$, a similar argument can be applied. Or merely observe that $u^{-1} = s^{-1}t^{-1}$ is a rotation by the previous case, and therefore so is u.

Finally, take the case of two rotations, $s = s(O, \alpha)$, $s' = s(O', \alpha')$. To compute the product $u = ss'$ when $O \neq O'$, define three lines in $\mathbb{R}^2$ as follows:

- l is the line through O and O',

- l' is the line through O such that the angle from l' to l is $\alpha/2$,

- l'' is the line through O' such that the angle from l to l'' is $\alpha'/2$.

Then by Theorem 4.1,

$$s = r(l')r(l), \quad s' = r(l)r(l''),$$

whence

$$u = ss' = r(l')r(l''),$$

as above. There are now two cases according as l' and l'' are parallel or not, and these are illustrated in Fig. 4.5. In the former case, the composite is a translation, and this occurs precisely when $\alpha/2$ and $\alpha'/2$ are supplementary angles or, equivalently, $\alpha' = -\alpha \bmod 2\pi$. Otherwise we are in the latter case, and the composite is the rotation $s(P, 2\phi)$, where $P = l' \cap l''$ and $2\phi = \alpha + \alpha'$.

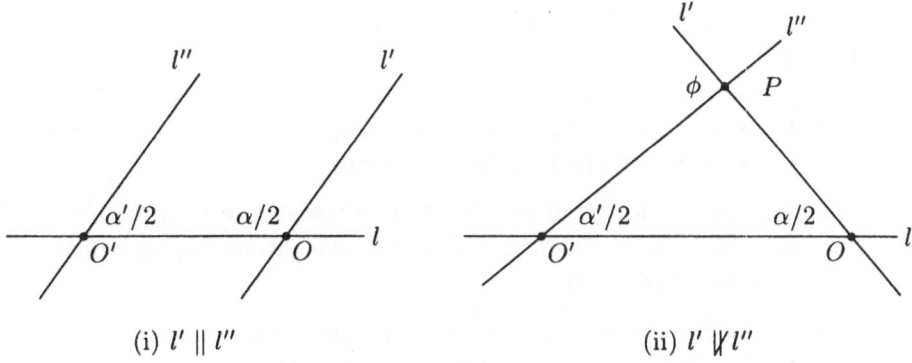

(i) $l' \parallel l''$ (ii) $l' \nparallel l''$

Figure 4.5   Composing rotations

To summarise, the product of four reflections in $\mathbb{E}$ is either a translation or a rotation, and as such is equal to the product of two reflections. It follows that when $n \geq 4$ any product of n reflections is a product of $n - 2$, and so, by an obvious induction, is a product of at most three. Thus all finite products of reflections are described by Theorems 4.1 and 4.2 above.

An appeal to the Normal Form Theorem shows that we have in fact proved a more general result. By that theorem together with Theorem 4.1, *every* isometry of $\mathbb{R}^2$ is a product of (at most five) reflections. Hence, all eventualities are catered for by Theorems 4.1 and 4.2, and we have proved the following definitive result.

Theorem 4.3

Every non-trivial isometry of $\mathbb{R}^2$ is the product of at most three reflections, and is either a rotation, translation, reflection or glide reflection. These four types are characterised in terms of fixed points and orientation as in the table below.

$\square$

Table 4.1   Characterisation of plane isometries

Fixed points?	Yes	No
OP	Rotation	Translation
OR	Reflection	Glide reflection

EXERCISES

4.1. Draw a diagram illustrating the result of composing the rotation $s = s(O, \alpha)$ and the translation $t = t(\mathbf{a})$.

4.2. Let r, s, t be a reflection, a rotation, a translation respectively. Find necessary and sufficient conditions on the parameters for (a) rs, (b) rt to be a reflection.

4.3. Prove that, in the decomposition of a glide reflection as the product rt of a reflection r and a translation t along the axis of r, the factors are unique.

4.4. Show that any glide reflection can be written as the product rs of a reflection r and a rotation s. Comment on the uniqueness of this decomposition.

4.5. Let p, q be the glide reflection determined by the ordered pairs

(P, P'), (Q, Q') respectively. Prove that their product pq is a translation if and only if $PP' \| QQ'$.

4.6. Show that any glide reflection can be written as the product of reflections in the sides of an equilateral triangle.

4.7. Prove that any rotation of $\mathbb{R}^3$ can be written as the product of two reflections. Deduce that the product of two rotations of $\mathbb{R}^3$ with intersecting axes is again a rotation. What if the axes are parallel?

4.8. Give another proof, along the lines of that of the Normal Form Theorem, that any isometry of $\mathbb{R}^2$ is the product of at most three reflections.

4.9. Extend the result of the previous exercise by showing that any isometry of $\mathbb{R}^3$ is the product of at most four reflections.

4.10. Prove that any OR isometry u of $\mathbb{R}^2$ is of the form rt, where t is a translation and r is reflection in some line through a prescribed point O.

5
Generators and Relations

Generators and relations provide an efficient means whereby a particular group may be defined or presented. We write

$$G = \langle X \mid R \rangle, \tag{5.1}$$

where the symbols X, R, G have the following meaning.

The set X of **generators** consists of symbols, usually finite in number, say $x_1, \ldots, x_n$, $n \in \mathbb{N} \cup \{0\}$. We think of the symbols $x_i^{\pm 1}$, $1 \leq i \leq n$, as letters in an alphabet $X^{\pm}$ from which **words** can be formed. The **length** of a word is the number of its letters, assumed to be finite, and we allow the empty word e of length zero. A word is **reduced** if it does not involve the letters $x_i^{\pm 1}$ in adjacent places for any i, $1 \leq i \leq n$, and the set of all reduced words is denoted by $F(X)$.

The set R of **defining relations** consists of relations, that is, equations between words, usually finite in number, say $u_i = v_i$, where $u_i, v_i \in F(X)$, $1 \leq i \leq m$, $m \in \mathbb{N} \cup \{0\}$.

Then we say that $\langle X \mid R \rangle$ is a **presentation** of a group G, that is, (5.1) holds, if the following three conditions are satisfied:

(i) X generates G in the usual sense: $X \subseteq G$ and $G = \langle X \rangle$, that is, every element of G can be written as a word in $X^{\pm}$,

(ii) the equations in R all hold in G, and

(iii) any equation between words in $X^{\pm}$ that holds in G is a consequence of the relations in R.

There is a lack of precision here insofar as the word "consequence" in condition (iii) has not been defined. Its meaning will be made intuitively clear in Section 5.1 below by means of examples and diagrams. The fastidious reader will be pleased to hear that there are various ways of supplying the missing rigour, such as the following rather unwieldy process.

Step 1. Prove that the set $F(X)$ of words in $X^{\pm}$ forms a group under concatenation plus cancellation. (The stumbling block here is, oddly enough, the associative law.) This is the **free group** on X, sometimes written $\langle X \mid \rangle$, of which an important invariant is the **rank**, $n = |X|$.

Step 2. Using the multiplication in $F(X)$, convert each equation $u_i = v_i$ to the equivalent form $r_i = 1$, where $r_i = u_i v_i^{-1}$, $1 \leq i \leq m$, then replace each defining relation in R by the corresponding **relator** r_i, $1 \leq i \leq m$.

Step 3. Now that $R \subseteq F(X)$, we can form its **normal closure** $\overline{R}$, which may be thought of either as the intersection of all normal subgroups of $F(X)$ that contain R, or as the set of all finite products of conjugates of elements of R and their inverses.

Step 4. Since $\overline{R} \lhd F(X)$, we can define $G = F(X)/\overline{R}$.

5.1 Examples

Let us quickly dispense with the trivial case $|X| = 0$, that is, X, and hence also R, is the empty set. Then the group $G = \langle \mid \rangle$ consists of the empty word e only. So $|G| = 1$ and G is the trivial group, also known as the free group of rank zero. So consider Example 5.1, where $|X| = n = 1, 2, 3$ in turn.

Example 5.1

As mentioned in Chapter 3, groups $G = \langle X \mid R \rangle$ with $|X| = 1$ are called cyclic groups. So let $X = \{x\}$ and observe that a reduced word in $X^{\pm}$ is just a power of x, positive, negative or zero:

$$F(X) = \{x^k \mid k \in \mathbb{Z}\}.$$

The multiplication in this group is given by the first formula in (3.5), and we recognise the infinite cyclic group

$$Z = \langle x \mid \rangle, \tag{5.2}$$

which is thus the free group of rank one.

Moving on to the case $|R| = 1$, we take the single relator $x^n \in F(X)$. Ignoring the trivial case $n = 0$ and observing that otherwise the relators x^n and x^{-n} are equivalent, we may assume that $n \in \mathbb{N}$. Then we get the cyclic group of order n,

$$Z_n = \langle x \mid x^n \rangle, \quad n \in \mathbb{N}, \tag{5.3}$$

studied in Section 3.2. In this group, x has order n, and so the statement at the end of Section 3.2 that for $m \in \mathbb{Z}$

$$x^m = 1 \Leftrightarrow m \text{ is a multiple of } n$$

merely asserts that the "consequences" of the defining relation $x^n = 1$ are precisely the relations $x^{kn} = 1$, $k \in \mathbb{Z}$. This ties in nicely with the fact that $\overline{R} = \{x^{kn} \mid k \in \mathbb{Z}\}$ in this case.

The case $|R| = 2$ reduces to the previous one in view of the fact that, for $m, n \in \mathbb{N}$,

$$x^k = x^l = 1 \Leftrightarrow x^h = 1,$$

where h is the highest common factor of k and l (exercise). By induction, any finite set of relators also reduces to the case $|R| = 1$. Moreover, since there is no such thing as an infinite strictly decreasing sequence of distinct positive integers, the iterative process of recursively incorporating a new integer into the highest common factor must stabilise after a finite number of steps. That settles the case when R is countably infinite and, since $F(X)$ is countable, there are no more cases. We have thus classified all cyclic groups into two types, (5.2) and (5.3) above, which agrees with the result in Section 3.2.

Example 5.2

We have had three examples so far with $|x| = 2$: formulae (1.16), (3.10) and (3.12). Translating into the new alphabet $X = \{x, y\}$ and replacing relations by relators, these become

$$D_\infty = \langle x, y \mid x^2, (xy)^2 \rangle, \tag{5.4}$$

$$D_{2n} = \langle x, y \mid y^n, x^2, (xy)^2 \rangle, \tag{5.5}$$

$$Z \times Z = \langle x, y \mid [x, y] \rangle, \tag{5.6}$$

respectively, where $n \in \mathbb{N}$ and $[x, y]$ denotes the commutator $x^{-1}y^{-1}xy$. Note that in the dihedral groups (5.4) and (5.5) we have

$$x^2 = 1 \Leftrightarrow x = x^{-1},$$

and so

$$(xy)^2 = 1 \Leftrightarrow x^{-1}yx = y^{-1}$$

in these groups. Note further that the relations

$$y^n = 1, \quad x^2 = 1, \quad y^x = y^{-1}$$

define the subgroup $\langle y \rangle$ as Z_n, the subgroup $\langle x \rangle$ as Z_2, the action ι of the latter on the former, respectively. This presentation thus displays in graphic form the structure of D_{2n} as the semidirect product $Z_2 \times_\iota Z_n$. We shall develop this idea in the next section and exploit it extensively in later chapters.

Similar remarks apply to the free abelian group of rank 2 presented in (5.6): each of y and x generates an infinite cyclic group, and the action of the latter on the former is given by

$$x^{-1}yx = x,$$

which is equivalent to the defining relation $[x, y] = 1$. We thus obtain the semidirect product of Z by Z with trivial action, namely, the direct product. It takes little imagination to guess (correctly) the general formula: if $G = \langle X \mid R \rangle$, $H = \langle Y \mid S \rangle$, then

$$G \times H = \langle X, Y \mid R, S, [X, Y] \rangle, \tag{5.7}$$

where $[X, Y]$ is the set of commutators $\{[x, y] \mid x \in X, y \in Y\}$. An easy induction now gives the standard presentation for the free abelian group of rank n:

$$Z^n = \langle x_i, 1 \le i \le n \mid [x_i, x_j], 1 \le i < j \le n \rangle. \tag{5.8}$$

These groups will reappear in Section 5.4 below.

Example 5.3

To gratify the thirst for novelty, here is an example with $n = 3$. Consider the group

$$Q = \langle i, j, k \mid ij = k, jk = i, ki = j \rangle,$$

in which the defining relations may possibly be familiar. We shall elucidate the structure of Q, picking up various interesting points along the way.

First replace i, j, k by x, y, z respectively to get what is clearly another presentation for the same group:

$$Q = \langle x, y, z \mid xy = z, yz = x, zx = y \rangle.$$

Next note that in the presence of the first relation the second and third are equivalent to $yxy = x$ and $xyx = y$ respectively, so we have

$$Q = \langle x, y, z \mid xy = z, yxy = x, xyx = y \rangle.$$

Now the generator z, which appears only once in the defining relations (as an abbreviation of xy), is superfluous:

$$Q = \langle x, y \mid yxy = x, xyx = y \rangle. \tag{5.9}$$

All this is more or less routine and will be formalised in Section 5.3 below. The next step is rather less obvious: we show that the last pair of defining relations has as a consequence the relation $y^4 = 1$. The proof is expressed in the form of the diagram in Fig. 5.1. The oriented edges are labelled by generators in such a way that the boundary label of each of the four quadrilaterals, read from some vertex in some direction, is a relator, where passage along an edge with or against the orientation (arrow) picks up the corresponding generator or its inverse, respectively. The deduction is that the boundary label of the whole diagram, namely y^4, is equal to 1 in Q. It is a simple exercise to read off from the diagram an expression for y^4 as a product of conjugates in $F(\langle x, y \rangle)$ of the relators $x^{-1}yxy$, $y^{-1}xyx$ and their inverses.

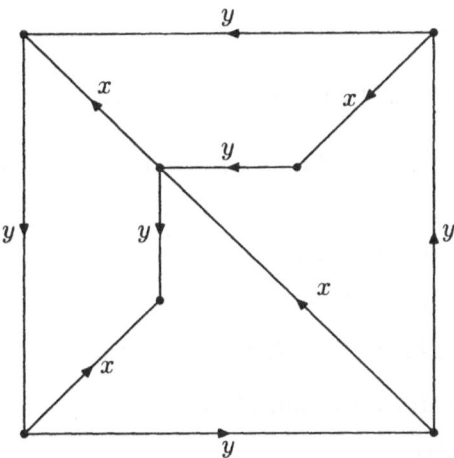

Figure 5.1   A van Kampen diagram

To finish the job, put $H = \langle y \rangle \leq Q$ so that $|H| \leq 4$. Next observe that the relation $yxy = x$ can be rewritten $x^{-1}yx = y^{-1}$ or $xyx^{-1} = y^{-1}$, whence H is a normal subgroup of Q. Finally, using both defining relations, we have

$$x^2 = x(yxy) = (xyx)y = y^2 \in H,$$

which proves that $|Q : H| \leq 2$. Thus $|Q| \leq 8$ by Lagrange's theorem.

To prove the reverse inequality, consider the matrices

$$X = \begin{pmatrix} 0 & 1 \\ -1 & 0 \end{pmatrix}, \quad Y = \begin{pmatrix} 0 & i \\ i & 0 \end{pmatrix}$$

over the complex field. Having verified that

$$XYX = Y, \quad YXY = X,$$

we deduce that the map

$$\rho: Q \to \langle X, Y \rangle, \quad x \mapsto X, \quad y \mapsto Y,$$

is a homomorphism into the group of invertible 2×2 matrices over $\mathbb{C}$ under multiplication. Having further verified that

$$|Y| = 4, \quad X \notin \langle Y \rangle,$$

we deduce from Lagrange's theorem that $\langle X, Y \rangle = \mathrm{Im}\, \rho$ has order at least 8, so that $|Q| \geq 8$ also.

We have shown that Q has exactly 8 elements. This is the famous quaternion group of order 8.

5.2 Semidirect Products Again

Recall the notation $G = K \times_\alpha H$ from Section 3.2, in which the three ingredients are as follows:

(i) H is a normal subgroup of G,

(ii) K is a complement for H in G,

(iii) α defines the action of K on H by conjugation.

The structure of G is then determined by

(i) the structure of H,

(ii) the structure of K,

(iii) the action α.

These three factors are encoded in a presentation of G by specifying

(i) a presentation $\langle Y \mid S \rangle$ of H,

(ii) a presentation $\langle X \mid R \rangle$ of K,

(iii) the conjugates $y^x = w(X)$, $x \in X$, $y \in Y$.

We express this by writing

$$G = \langle X, Y \mid R = 1, S = 1, Y^X = W(X) \rangle, \tag{5.10}$$

a presentation with $|X| + |Y|$ generators and $|R| + |S| + |X|\,|Y|$ defining relations (where $W(X)$ denotes the set of words $w(X)$ in $X^{\pm 1}$, one for each $x \in X$, $y \in Y$).

Caveat. What we are saying here is that every semidirect product of $H = \langle Y \mid S \rangle$ by $K = \langle X \mid R \rangle$ has a presentation of the form (5.10). It is not true that any such presentation determines such a semidirect product. The point is that the words $w(X)$ in $X^{\pm}$ are not arbitrary, but are determined by the homomorphism $\alpha \colon K \to \text{Aut}(H)$. See Exercise 5.9 for an illustration.

Rather than attempting to make this informal derivation rigorous, which would carry us too far afield, we shall study a few illuminating examples. The dihedral groups have already appeared in (5.4) and (5.5), and (5.6), (5.7) and (5.8) illustrate the special case of direct products. Our next example is a little more ambitious and requires a few preliminaries.

Recall from Section 1.2 that the symmetric group S_n of order $n!$ is defined as the group of all permutations of the set $\Omega_n = \{1, 2, \ldots, n\}$ of the first n positive integers under composition of maps. Here is a brief summary of the elementary theory of these groups.

A permutation $\sigma \in S_n$ is called a **cycle** of length l, or l-cycle, if there are distinct elements $a_1, a_2, \ldots, a_l \in \Omega_n$ such that

(a) $a_k \sigma = a_{k+1}$, $1 \le k < l$,

(b) $a_l \sigma = a_1$,

(c) σ fixes the remaining $n - l$ elements of Ω_n,

and then we write $\sigma = (a_1 a_2 \cdots a_l)$. 2-cycles are called **transpositions**, and 1-cycles are often ignored. One readily shows that

 (i) any $\sigma \in S_n$ is a product of cycles,

 (ii) any cycle is a product of transpositions, and

(iii) any transposition is a product of adjacent transpositions, $(i \ i+1)$, $1 \le i < n$.

In (i), the cycles can be chosen disjoint, that is, no $a \in \Omega_n$ is moved by two of them, and then the decomposition is more or less unique. On the other hand, the decomposition in (ii) is far from unique. The $n - 1$ adjacent transpositions in (iii) provide the first step in getting a presentation for S_n; we shall not go so far.

Focusing on (ii), a permutation is said to be **even** or **odd** according as it is the product of an even or odd number of transpositions. It is easy to see that

- the product of two even permutations,

- the identity permutation, and

- the inverse of an even permutation

are all even. What is not easy to see is that the concept of parity (even-or-oddness) is well defined: it may be that a permutation has two decompositions, into an even and odd number of transpositions respectively. It turns out that this cannot happen, and here is a (bogus) proof.

Regard S_n as a subgroup of Isom($\mathbb{R}^n$) by letting its elements act as permutations of the coordinates of a vector $\mathbf{x} = (x_1, x_2, \ldots, x_n)$ in the obvious way. Thus, for example, (ij) transposes the coordinates x_i, x_j and fixes all the others: this is just reflection in the hyperplane $x_i = x_j$ and as such is OR. The usual rules imply that a permutation expressible as both an odd and an even number of transpositions is both OP and OR. But this is impossible, so there aren't any.

Parity is nevertheless well defined, and so the even permutations do indeed form a subgroup of S_n, called the **alternating group** of degree n and denoted by A_n. Since multiplication by the transposition (12) is obviously a bijection from A_n to $S_n \setminus A_n$, it follows that $|S_n : A_n| = 2$, and so A_n is a normal subgroup of S_n when $n \geq 2$. A classical result of Galois asserts that A_n is the only proper non-trivial normal subgroup of S_n when $n \geq 5$, and it is this fact that underlies the proof that the general polynomial equation of degree at least 5 is not soluble by radicals. In particular, A_5 is a **simple** group, that is, it has no proper non-trivial normal subgroup. Because of difficulties arising from this fact, we limit ourselves for the moment to finding presentations for S_n and A_n when $n \leq 4$.

Example 5.4

The groups S_1 and A_1 are both trivial and so is A_2. S_2 is cyclic of order 2 and A_3 is cyclic of order 3. S_3 is isomorphic to D_6, which is given by (5.5) with $n = 3$. So consider A_4, whose elements are

$$1, (12)(34), (13)(24), (14)(23), (123), (132),$$
$$(124), (142), (134), (143), (234), (243).$$

Putting

$$a = (12)(34), \quad b = (13)(24), \quad c = (14)(23),$$

we see that

$$a^2 = b^2 = c^2 = 1, \quad ab = ba = c.$$

Thus, $V = \langle a, b \rangle$ is a subgroup of A_4 isomorphic to $Z_2 \times Z_2$. Putting $x = (132)$, we see that

$$a^x = b, \quad b^x = c, \quad c^x = a. \tag{5.11}$$

Since $x \notin V$, we have $A_4 = \langle a, b, x \rangle$, whence V is a normal subgroup of A_4 with complement $K = \langle x \rangle$. Thus, $A_4 = K \times_\alpha V$, where α is given by (5.11), and we obtain

$$A_4 = \langle a, b, x \mid a^2 = b^2 = (ab)^2 = 1, x^3 = 1, a^x = b, b^x = ab \rangle \qquad (5.12)$$

as an example of (5.10).

To get a presentation for S_4, let $y = (14)$ act on A_4 in the natural way, so that

$$y^2 = 1, \quad a^y = b, \quad b^y = a, \quad x^y = bx^{-1}.$$

Thus, according to (5.10), S_4 is generated by a, b, x, y and defined by the relations

$$a^2 = b^2 = (ab)^2 = 1 = x^3, \quad a^x = b, \quad b^x = ab,$$
$$y^2 = 1, \quad a^y = b, \quad b^y = a, \quad x^y = bx^{-1}. \qquad (5.13)$$

Notice in passing that, by taking $z = (12)$ in place of y here, the resulting presentation exhibits S_4 as a semidirect product of $\langle x, z \rangle \cong S_3$ and $V = \langle a, b \rangle$. The action is very natural: $\mathrm{Aut}(Z_2 \times Z_2) \cong S_3$ (see Exercise 5.10).

These rather unweidly presentations for A_4 and S_4 will be simplified in the next section.

Example 5.5

This is another example of compound type: we take a group and extend it twice. Consider the free abelian group of rank 2

$$G_1 = Z^2 = \langle a, b \mid ab = ba \rangle$$

and the automorphism $s \in \mathrm{Aut}(Z^2)$ given by

$$a^s = b, \quad b^s = a^{-1}.$$

The effect of s on an arbitrary element of Z^2 is thus

$$(a^k b^l)^s = a^{-l} b^k, \quad k, l \in \mathbb{Z}.$$

Note in passing that it is often useful to think of Z^2 as the additive group $\mathbb{Z}^2$ via the correspondence $a^k b^l \to (k, l)$. Then the generators a, b look very much like the basis vectors $(1, 0)$, $(0, 1)$ of a vector space. Homomorphisms then resemble linear transformations, and our automorphism s can be written as a matrix, coincidentally the matrix

$$X = \begin{pmatrix} 0 & 1 \\ -1 & 0 \end{pmatrix}$$

which appeared in Section 5.1, where we saw that $|X| = 4$. The fact that $|s| = 4$ can also be seen by applying it repeatedly to a:

$$a \mapsto b \mapsto a^{-1} \mapsto b^{-1} \mapsto a \mapsto b.$$

This shows that s^4 fixes both a and b, and hence the whole of G_1. Using (5.10), the resulting semidirect product of Z_4 and G_1 has the presentation

$$G_4 = \langle a, b, s \mid ab = ba, s^4 = 1, a^s = b, b^s = a^{-1} \rangle.$$

To extend this group, take $r \in \text{Aut}(G_4)$ given by

$$a^r = a, \quad b^r = b^{-1}, \quad s^r = s^{-1}.$$

Note that the relations defining G_4 pass under the action of r to equivalent ones:

$$ab^{-1} = b^{-1}a, \quad s^{-4} = 1, \quad a^{s^{-1}} = b^{-1}, \quad (b^{-1})^{s^{-1}} = a^{-1},$$

whence r is well defined. The obvious fact that $r^2 = 1$ guarantees that r is a bijection. We again use (5.10) to assemble a presentation for the corresponding semidirect product,

$$G_4^1 = \langle a, b, s, r | ab = ba, s^4 = 1, a^s = b, b^s = a^{-1},$$
$$r^2 = 1, a^r = a, b^r = b^{-1}, s^r = s^{-1} \rangle.$$

Again notice that a repartitioning of the relations shows in accordance with (5.10) that G_4^1 is a semidirect product of $G_1 = \langle a, b \rangle$ by $\langle s, r \rangle = D_8$.

This group has a nice geometrical realization. Consider the lattice $\mathbb{Z}^2$ of integer points in $\mathbb{R}^2$ as depicted in Fig. 5.2, where O, A, B denote the points $(0,0)$, $(1,0)$, $(0,1)$ respectively and l is the line through O, A. A little thought shows that the isometries

$$a = t(1,0), \quad b = t(0,1), \quad s = s(O, \pi/2), \quad r = r(l)$$

satisfy the eight relations defining G_4^1. A little more thought shows that $G_4^1 = \text{Sym}(\mathbb{Z}^2)$.

The groups G_1, G_4, G_4^1 will reappear in Chapters 7–9 along with fourteen companions.

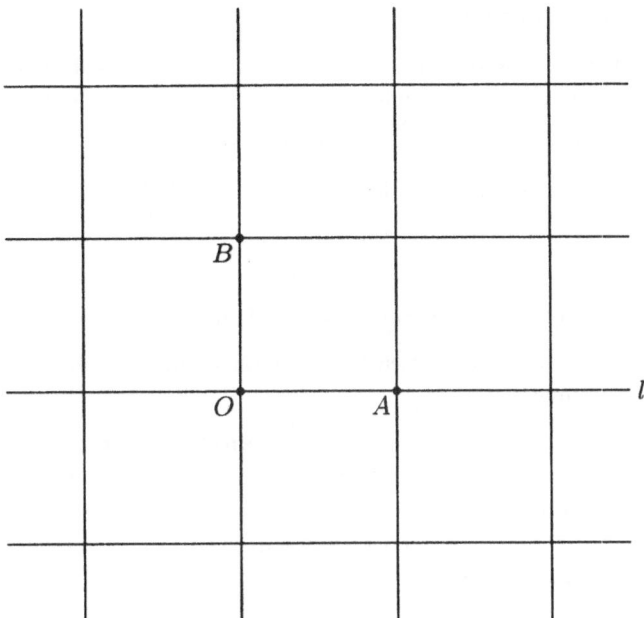

Figure 5.2   The lattice $\mathbb{Z}^2$ in $\mathbb{R}^2$

5.3 Change of Presentation

We shall discuss the problems of existence and uniqueness of presentations. The former is easy: every group has a presentation. To see this, let G be any group and denote its binary operation by $\mu\colon G \times G \to G$. Then take

$$X = G, \quad R = \{xy = (x, y)\mu \mid x, y \in G\}.$$

Even a formal proof that $G = \langle X \mid R \rangle$ is not hard to come by: the relations R define the multiplication table of G, which in turn defines the group. Of course in general, and in what follows, we look for slicker presentations than this. Note that we have shown that every finite group has a finite presentation, that is, both X and R are finite sets. On the question of uniqueness, to which we now turn, we restrict our attention to finite presentations.

It is already clear from the study of the quaternion group Q in Example 5.3 above that a group can have several presentations. The four presentations of the group Q given there show that we can

(a) relabel (consistently) the generators,

(b) replace relations by equivalent ones, and even

(c) lose generators, under certain conditions

and come up with a new presentation of the same group. All of these transformations can be manufactured from the following four elementary types.

$R+$: insert a relation that is a consequence of the existing ones,

$X+$: insert a new generator, and a new relation defining it in terms of the existing ones,

and their inverses,

R_-: delete a relation that is a consequence of the others,

$X-$: delete a generator that appears only once in the relations, along with the relation involving it.

These are the four elementary **Tietze transformations**, and it is pretty clear (and not hard to prove) that their application to a presentation yields another presentation of the same group. The same obviously goes for the composite of a finite sequence of such transformations. The converse, though not so clear, is even easier to prove.

Theorem 5.1

If $\langle X \mid R(X) \rangle$ and $\langle Y \mid S(Y) \rangle$ are two finite presentations of the same group G, then one can be obtained from the other by a finite sequence of elementary Tietze transformations.

Proof

Since the set X generates G, every element of Y can be written as a word in $X^{\pm}$; we abbreviate these $|Y|$ equations to $Y = Y(X)$. By symmetry, we have $X = X(Y)$. The rest of the proof takes the form of an algorithm with input $X \mid R(X)$, output $Y \mid S(Y)$, and each step the application of a block of elementary Tietze transformations.

0. input $= X \mid R(X)$
1. $X+$: $X, Y \mid R(X), Y = Y(X)$
2. $R+$: $X, Y \mid R(X), Y = Y(X), X = X(Y)$
3. $R+$: $X, Y \mid R(X), Y = Y(X), X = X(Y), R(X(Y)), Y = Y(X(Y))$
4. $R-$: $X, Y \mid X = X(Y), R(X(Y)), Y = Y(X(Y))$
5. $X-$: $Y \mid R(X(Y)), Y = Y(X(Y))$ (5.14)
6. $R+$: $Y \mid R(X(Y)), Y = Y(X(Y)), S(Y)$
7. $R-$: $Y \mid S(Y) =$ output. $\square$ (5.15)

Remark 5.1

Steps 6 and 7 are valid because the relations $S(Y)$ hold in and define G respectively.

Remark 5.2

The presentation (5.14) defines G on the new generators Y. The relations $R(X(Y))$ are obtained by substitution, and both the meaning of and the need for the equations $Y = Y(X(Y))$ are illustrated by the following simple example. The group $Z_3 = \langle x \mid x^3 = 1 \rangle$ is generated by the element $y = x^2$, since $x = y^2$. The relations $R(X(Y))$, $Y = Y(X(Y))$ then read

$$(y^2)^3 = 1, \quad y = (y^2)^2,$$

which define Z_3 on the generator y.

Remark 5.3

Despite superficial appearances to the contrary, this algorithm does not solve the problem of deciding whether or not two finite presentations $\langle X \mid R \rangle$, $\langle Y \mid S \rangle$ define the same group. This is the Isomorphism Problem, the hardest of three decision problems recognized by Max Dehn in 1912. The easiest is the Word Problem, which is equivalent to the special case $X = Y$. The third is the Conjugacy Problem, and all three are insoluble.

On a more positive note, let us illustrate the use of Tietze transformations by continuing with Example 5.4 from Section 5.2.

Example 5.6

First take the presentation (5.12) of A_4. Of the six relations

$$a^2 = b^2 = (ab)^2 = 1 = x^3, \quad a^x = b, \quad b^x = ab,$$

the fifth shows that the generator b is superfluous. Substituting it out ($R+$ and $R-$) and applying $X-$, we get

$$a^2 = (a^x)^2 = (aa^x)^2 = 1 = x^3, \quad a^{x^2} = aa^x.$$

The second of these is now redundant and, using the fifth, so is the third. Since $a^2 = 1 = x^3$, the last relation can be written

$$1 = a^{x^2} a^x a = xaxaxa,$$

and we finish up with

$$A_4 = \langle a, x \mid a^2 = x^3 = (xa)^3 = 1 \rangle. \qquad (5.16)$$

Turning now to S_4, defined by the relations (5.13), we first eliminate $b = a^x$ as above to get defining relations

$$a^2 = x^3 = (xa)^3 = 1, \quad y^2 = 1, \quad a^y = a^x, \quad a^{xy} = a, \quad x^y = x^{-1}a. \qquad (5.17)$$

Now use the last relation to eliminate $a = (xy)^2$:

$$(xy)^4 = x^3 = (x(xy)^2)^2 = 1 = y^2, \quad ((xy)^2)^y = ((xy)^2)^x, \quad ((xy)^2)^{xy} = (xy)^2.$$

Since $y^2 = 1$, the last two relations are equivalent and since they merely assert that xy commutes with $(xy)^2$, they are redundant. To see that the third relation, $(x(xy)^2)^3 = 1$, is a consequence of the remaining three, observe that the first relation, $(xy)^4 = 1$, can be written

$$xyxyx = y^{-1}x^{-1}y^{-1} = yx^2y$$

in the presence of $x^3 = 1 = y^2$. Then we have

$$\begin{aligned}
(x(xy)^2)^3 &= x \cdot xyxyx \cdot xyxyx \cdot xyxy \\
&= x \cdot yx^2y \cdot yx^2y \cdot xyxy \\
&= xyx^4yxyxy, \quad \text{as } y^2 = 1, \\
&= (xy)^4, \quad \text{as } x^3 = 1, \\
&= 1, \quad \text{as } (xy)^4 = 1.
\end{aligned}$$

We therefore finish up with the sleek presentation

$$S_4 = \langle x, y \mid y^2 = x^3 = (xy)^4 = 1 \rangle. \qquad (5.18)$$

Here endeth Example 5.6.

Two salient features of the presentations derived above for the groups S_3, A_4 and S_4 are worthy of note:

(a) the rather striking uniformity they exhibit, and

(b) the clumsiness of the calculations used to get them.

As to (b), there are at least two other ways of establishing these presentations. The first is an elegant and very general method for handling finite presentations by systematic enumeration of cosets. This method (not quite an algorithm) is a pillar of Computational Group Theory, and as such is sadly beyond our scope. The second method is more specialized and exploits geometrically the uniformity referred to in (a). This link between algebra and geometry is described in the following brief but important section.

5.4 Triangle Groups

Consider the sequence of groups given by the presentations

$$\Delta_n^+ = \langle a, b \mid a^2 = b^3 = (ab)^n = 1 \rangle. \tag{5.19}$$

Ignoring the trivial case $n = 1$, it follows from Examples 5.4 and 5.6 above that when $n = 2, 3, 4$, $\Delta_n^+ \cong S_3, A_4, S_4$, respectively, and one is tempted to look for a pattern. In the algebraic sense, such hopes

(a) diminish at $n = 5$: it turns out that $\Delta_5^+ \cong A_5$,

(b) suffer a severe setback at $n = 6$: Δ_6^+ is an infinite group,

(c) are destroyed completely at $n = 7$: Δ_7^+ contains a copy of the free group of rank 2.

This unpleasant but intriguing behaviour is, of course, a consequence of the geometry underlying these groups, which runs as follows.

Let x, y, z denote reflections in the three sides of a triangle. Then, in accordance with Chapter 4,

$$a = xy, \quad b = yz, \quad c = zx$$

are rotations about the three vertices. In the case when the angles in the triangle are all submultiples of π, say π/l, π/m, π/n, $l, m, n \in \mathbb{N}$, the rotations a, b, c all have finite order:

$$a^l = b^m = c^n = 1$$

(recall that the angle of rotation is twice the angle between the axes of reflection). Observing that, as $x^2 = y^2 = z^2 = 1$,

$$c = zx = zy \cdot yx = b^{-1}a^{-1} = (ab)^{-1},$$

we get the relations

$$a^l = b^m = (ab)^n = 1,$$

of which the defining relations in (5.19) form the special case $(l, m) = (2, 3)$.

This motivates the definition of the **triangle group**

$$\Delta(l, m, n) = \langle x, y, z \mid x^2 = y^2 = z^2 = 1, (xy)^l = (yz)^m = (zx)^n = 1 \rangle, \tag{5.20}$$

and its OP subgroup

$$\Delta^+(l, m, n) = \langle a, b \mid a^l = b^m = (ab)^n = 1 \rangle. \tag{5.21}$$

It is easy to show (Exercise 5.13) that Δ is the semidirect product $\Delta^+ \times_\alpha Z_2$, where α is the automorphism of Δ^+ inverting a and b.

It should be pretty clear that the structure of the group $\Delta(l, m, n)$ will depend on the triangle from which it springs, or rather on the geometry of the space in which that triangle lies. For example, if that space is a metric space M on which the elements of $\Delta(l, m, n)$ act as isometries, the orbit of the triangle under the action of this group (that is, the set of its images) will form a **tessellation** (or tiling) of M, and the order of the group will be equal to the number of tiles. So what does this M look like?

The happy fact is that the answer depends on one and only one parameter, the angle-sum of the triangle,

$$\sigma = \pi/l + \pi/m + \pi/n. \qquad (5.22)$$

To be specific, the geometry is determined by the trichotomy

$$\sigma > \pi, \quad \sigma = \pi, \quad \sigma < \pi.$$

In the case of equality, as when $n = 6$ in (5.19), we take the familiar Euclidean plane $\mathbb{R}^2$. When $\sigma > \pi$ ($2 \leq n \leq 5$ in (5.19)), the appropriate space is the sphere $\mathbb{S}^2$ studied in Chapter 10 below. In all remaining cases we have $\sigma < \pi$ ($n \geq 7$ in (5.19)), and the triangle can be drawn in the hyperbolic plane $\mathbb{H}^2$, first studied in the century before last by Bolyai, Gauss and Lobachevskii. This will form the setting for Chapter 11.

5.5 Abelian Groups

In contrast to the triangle groups of the previous section, abelian groups are relatively well behaved. In fact, as already mentioned, there is a complete classification in the finitely generated case. We refer to this as the Basis Theorem; it is probably the most important theorem in this book and was first proved by Kronecker. The purpose of this section is to give an outline of the proof in the context of group presentations. We shall proceed in a number of steps, bearing in mind that any direct product of cyclic groups is abelian, in particular the direct product of r copies of the infinite cyclic group Z, which is written Z^r and called the **free abelian group of rank** r. What follows is a paraphrase, and the reader who gets stuck is advised to refer for clarification to the example at the end.

1. Abelianisation. We begin with an arbitrary finite presentation

$$G = \langle x_1, \ldots, x_n \mid r_1, \ldots, r_m \rangle, \quad m, n \in \mathbb{N}.$$

It is not hard to show that by throwing in the set

$$C = \{[x_i, x_j] \mid 1 \leq i < j \leq n\}$$

of commutators we get a presentation of G/G'. This new presentation can also be viewed as having been obtained from

$$A = \langle x_1, \ldots, x_n \mid C \rangle$$

by adjoining the relators $r_1, \ldots, r_m$. Now A is just the free abelian group of rank n, and so every n-generator abelian group is a quotient A/B of A. We now invoke a theorem of Dedekind which asserts that any subgroup of A is free abelian, of rank at most n. It follows that *every* finitely generated abelian group is finitely presented and thus of the form

$$\langle x_1, \ldots, x_n \mid r_1, \ldots, r_m, C \rangle.$$

2. The relation matrix. In deference to the fact that abelian groups are commutative, we shall adopt for the rest of this section the convention that the group operation is denoted by addition: we write $a + b$ in place of ab. In line with this, we write 0, $-a$, na ($n \in \mathbb{Z}$) in place of 1, a^{-1}, a^n. Then the images $y_1, \ldots, y_m$ of the relators $r_1, \ldots, r_m$ in the abelian group A take the form

$$y_i = \sum_{j=1}^{n} e_{ij} x_j, \quad 1 \le i \le m,$$

where $e_{ij} \in \mathbb{Z}$ is the exponent-sum of x_j in r_i. Since $B = \langle y_1, \ldots, y_m \rangle \le A$ is such that $A/B \cong G/G'$, it follows that the structure of G/G', and hence of any finitely generated abelian group, is determined by the corresponding **relation matrix** $E = (e_{ij})$.

3. Row and column operations. What follows in this step and the next will look very much like a piece of linear algebra. The idea is to reduce E to some canonical form by means of row and column operations that do not change the abelian group to which the matrix corresponds. However, since the "scalars" here are integers rather than elements of a field, the usual row operations require a slight modification, as follows:

P: transposing two rows, corresponding to a permutation of the y_i,

M: multiplying a row by -1, which inverts a y_i,

A: adding an integer multiple of one row to another, replacing y_i by $y_i + l y_k$, $k \neq i, l \in \mathbb{Z}$,

while keeping the other rows fixed, and likewise for columns (with corresponding effects on the x_j). A little thought shows that none of these operations changes the corresponding abelian group.

4. Smith normal form. The desired canonical form is an $m \times n$ matrix D with entries $d_1, \ldots, d_s$, where $s = \min(m, n)$, on the main diagonal and zeros elsewhere, where the d_i are non-negative integers and each divides the next. Note that the last condition implies that any 1s among the d_i occur at the beginning, and any 0s at the end. The question is, to what abelian group does such a matrix correspond? On (new) generators $x_1, \ldots, x_n$, we have relations $y_i = d_i x_i$, which correspond to $r_i = x_i^{d_i}$, $1 \le i \le s$. Therefore the presentation is, in multiplicative notation,

$$\langle x_1, \ldots, x_n \mid x_1^{d_1}, \ldots, x_s^{d_s}, C \rangle,$$

which is just the direct product

$$Z_{d_1} \times \cdots \times Z_{d_s} \times Z^r,$$

where Z^r is the free abelian group of rank $r = n - s$. Each d_i divides the next, and any Z_1s (trivial groups, at the beginning) are thrown away while any Z_0s (infinite cyclic groups, at the end) are absorbed into the Z^r.

5. *The algorithm.* This mimics the familiar Gauss–Jordan reduction of a matrix over a field to row-echelon form and is adapted to

(a) the problem: we actually want a *diagonal* normal form,

(b) our situation: the integers do *not* form a field.

As a result, the pivots

(a) are used to clear rows as well as columns, and columns of zeros are moved to the right, and

(b) may be integers greater than one.

Because of (a), the pivots all lie on the main diagonal, and the divisibility condition (each divides the next) is assured by the fact that, at each stage, the pivotal entry d becomes the highest common factor of all entries in rows and columns as yet untreated. The position of d_1 (top left-hand corner) is effected by **P**-operations, and its sign (positive) by **M**-operations. Finally, the divisibility condition is ensured by applying, via **A**-operations, Euclid's algorithm to pairs of elements in the same row or column.

6. *The Basis Theorem.* Every finitely generated abelian group is uniquely of the form
$$Z_{d_1} \times \cdots \times Z_{d_s} \times Z^r, \tag{5.23}$$

where r is a non-negative integer (called the **rank**) and the d_i are integers ≥ 2 (called the **invariant factors**) such that $d_i \mid d_{i+1}$, $1 \le i \le s - 1$.

7. *Uniqueness.* Let G be an abelian group of the form (5.23) and admitting a rival decomposition
$$Z_{e_1} \times \cdots \times Z_{e_t} \times Z^q$$

with like constraints on the parameters. An easy exercise shows that the elements of finite order in G form a subgroup, the **torsion subgroup** $\tau(G)$. We thus have alternative decompositions for $\tau(G)$,

$$Z_{d_1} \times \cdots \times Z_{d_s} \cong Z_{e_1} \times \cdots \times Z_{l_t}, \tag{5.24}$$

and $G/\tau(G)$,

$$Z^r \cong Z^q, \tag{5.25}$$

and we treat these separately, as follows.

First note that, for any abelian group A and any $k \in \mathbb{N}$, the map

$$\mu_k : A \to A, \quad a \mapsto ka, \tag{5.26}$$

of multiplication by k is a homomorphism (see Exercise 5.17). By considering the quotient by the image of μ_2 applied to the group in (5.25), we get

$$Z^r \cong Z^q \Rightarrow Z^r_2 \cong Z^q_2 \Rightarrow 2^r = 2^q \Rightarrow r = q.$$

Now apply μ_{d_1} to the group in (5.24). By Exercise 5.17,

$$\prod_{i=1}^{s} (d_i, d_1) = \prod_{j=1}^{t} (e_j, d_1). \tag{5.27}$$

The divisibility condition on the d_i ensures that every term on the left-hand side is equal to d_1. Since every term on the right is a divisor of, and hence at most, d_1, we deduce that $t \geq s$. By symmetry (replace d_1 by e_1), $s \leq t$. Hence, $s = t$ and then (5.26) implies that d_1 divides all e_j. In particular, d_1 divides e_1 and symmetry (again) gives $d_1 = e_1$. Factoring out Im μ_{d_1} and using induction, we get $e_i = d_i$, $1 \leq i \leq s$, as required.

8. Example. Let G be the group with presentation

$$\langle x_1, x_2, x_3, x_4, x_5 \mid r_1, r_2, r_3, r_4 \rangle,$$

where

$$r_1 = (x_1 x_4)^3 (x_2 x_5)^5 x_3^{-2}, \quad r_2 = (x_1 x_4)^5 (x_2 x_5)^9 x_3^{-4},$$
$$r_3 = (x_1 x_2 x_4 x_5)^{-3} (x_2 x_3^{-1} x_4)^{-6}, \quad r_4 = (x_1 x_2 x_4 x_5)^4.$$

To abelianise, throw in the commutators and convert to additive notation. Collecting terms, the generators of B are

$$y_1 = 3x_1 + 5x_2 - 2x_3 + 3x_4 + 5x_5,$$
$$y_2 = 5x_1 + 9x_2 - 4x_3 + 5x_4 + 9x_5,$$
$$y_3 = -3x_1 - 9x_2 + 6x_3 - 9x_4 - 3x_5,$$
$$y_4 = 4x_1 + 4x_2 + 4x_4 + 4x_5.$$

The relation matrix is thus

$$E = \begin{bmatrix} 3 & 5 & -2 & 3 & 5 \\ 5 & 9 & -4 & 5 & 9 \\ -3 & -9 & 6 & -9 & -3 \\ 4 & 4 & 0 & 4 & 4 \end{bmatrix}.$$

To apply the Euclidean algorithm to rows 1 and 2, subtract row 1 from row 2, then (new) row 2 from (old) row 1. This gives

$$\begin{bmatrix} 1 & 1 & 0 & 1 & 1 \\ 2 & 4 & -2 & 2 & 4 \\ -3 & -9 & 6 & -9 & -3 \\ 4 & 4 & 0 & 4 & 4 \end{bmatrix}.$$

Clear the first column to get

$$\begin{bmatrix} 1 & 1 & 0 & 1 & 1 \\ 0 & 2 & -2 & 0 & 2 \\ 0 & -6 & 6 & -6 & 0 \\ 0 & 0 & 0 & 0 & 0 \end{bmatrix}.$$

We thus have $d_1 = 1$. Clear the first row, delete the last, and focus on the new pivot 2 in what remains:

$$\begin{bmatrix} 2 & -2 & 0 & 2 \\ -6 & 6 & -6 & 0 \end{bmatrix} \rightarrow \begin{bmatrix} 2 & -2 & 0 & 2 \\ 0 & 0 & -6 & 6 \end{bmatrix} \rightarrow \begin{bmatrix} 2 & 0 & 0 & 0 \\ 0 & 0 & -6 & 6 \end{bmatrix},$$

clearing the first column and row in turn. So $d_2 = 2$ and the next pivot is 6:

$$\begin{bmatrix} 0 & -6 & 6 \end{bmatrix} \rightarrow \begin{bmatrix} 6 & -6 & 0 \end{bmatrix} \rightarrow \begin{bmatrix} 6 & 0 & 0 \end{bmatrix}.$$

Hence, the relation matrix E has Smith normal form

$$D = \begin{bmatrix} 1 & 0 & 0 & 0 & 0 \\ 0 & 2 & 0 & 0 & 0 \\ 0 & 0 & 6 & 0 & 0 \\ 0 & 0 & 0 & 0 & 0 \end{bmatrix},$$

so that $d_1 = 1$, $d_2 = 2$, $d_3 = 6$, $d_4 = 0$. Finally, ignoring d_1 and absorbing d_4, we get

$$G/G' \cong Z_2 \times Z_6 \times Z^2.$$

EXERCISES

5.1. Let G be any group and S any subset of G. Denote by S^G the set of all conjugates $g^{-1}sg$, where $s \in S$, $g \in G$. Prove that the subgroup $\overline{S} = \langle S^G \rangle$ coincides with the intersection I of all normal subgroups of G that contain S.

5.2. Show that, for an element x of a group G,

$$x^k = x^l = 1 \Leftrightarrow x^h = 1,$$

where h is the highest common factor (k, l) of integers k, l.

5.3. Consider the words y^4, $x^{-1}yxy$, $y^{-1}xyx \in F(\langle x, y \rangle)$. Use Figure 5.1 to express the first of these as a product of conjugates of the other two.

5.4. Check that the matrices $X = \begin{pmatrix} 0 & 1 \\ -1 & 0 \end{pmatrix}$, $Y = \begin{pmatrix} 0 & i \\ i & 0 \end{pmatrix}$ satisfy the relations $XYX = Y$, $YXY = X$. Show that $|Y| = 4$, $X \notin \langle Y \rangle$.

5.5. Show that the group

$$G = \langle x, y \mid x^3 = 1, y^5 = 1, y^x = y^2 \rangle$$

is *not* a semidirect product of Z_5 by Z_3.

5.6. Given a permutation $\sigma \in S_n = \text{Sym}(\Omega_n)$, define a relation on Ω_n as follows:

$$i \sim j \Leftrightarrow \exists k \in \mathbb{N} \; i\sigma^k = j.$$

Prove that this is an equivalence relation. Describe how the $\sim$-classes (or σ-**orbits**) can be used to express σ as a product of disjoint cycles in S_n. Comment on the uniqueness of this decomposition.

5.7. Express the cycle $\sigma = (a_1 a_2 \cdots a_l)$ as a product of $l - 1$ transpositions. Express any transposition $(ij) \in S_n$, $1 < i < j \le n$, in terms of the transpositions $(1i)$ and $(1j)$. Finally, express any $(1j)$ in terms of adjacent transpositions $(i \; i+1)$, $1 \le i < n$.

5.8. Prove that S_3 has the presentation

$$\langle x, y \mid y^3 = x^2 = (xy)^2 = 1 \rangle.$$

Show that two relations, $y^3 = x^2$ and $y^x = y^2$, suffice to define the group.

5.9. Show that the **metacyclic** group

$$G = \langle x, y \mid x^n = y^m = 1, y^x = y^r \rangle,$$

$n, m, r \in \mathbb{N}$, has order nh, where $h = (m, r^n - 1)$.

5.10. By using $z = (12)$ in place of $y = (14)$ in the last step in Example 5.4, exhibit S_4 as a semidirect product of S_3 and $Z_2 \times Z_2$.

5.11. Check that under the action in (5.17) on the group (5.16),

    (a) y sends each defining relation to a relation, and

    (b) y^2 sends each generator to itself.

5.12. Show that the presentation Δ_1^+ defines the trivial group.

5.13. Show that the map α with $a\alpha = a^{-1}$, $b\alpha = b^{-1}$ defines an automorphism of the group $\Delta^+(l, m, n)$ presented in (5.21). Prove that the resulting semidirect product with Z_2 is the group $\Delta(l, m, n)$ presented in (5.20).

5.14. Find all three solutions of the equation

$$1/l + 1/m + 1/n = 1$$

in integers l, m, n with $2 \leq l \leq m \leq n$. In each case, draw the corresponding triangle in $\mathbb{R}^2$ and (part of) the tessellation it generates.

5.15. Find all three solutions of the inequality

$$1/l + 1/m + 1/n > 1$$

in integers l, m, n with $2 \leq l \leq m \leq n$. In each case, try to imagine the corresponding triangle in $\mathbb{S}^2$ and the tessellation it generates.

5.16. Check that the set $\tau(G)$ of all torsion elements in an abelian group G forms a subgroup. Give an example to demonstrate the failure of the corresponding assertion for non-abelian groups.

5.17. Check that the map $\mu_k \colon A \to A$ of (5.26) is a homomorphism. Letting

$$\#_k(A) = |\mathrm{Ker}\, \mu_k| = |\{a \in A \mid ka = 0\}|,$$

prove that

$$\#_k(A \times B) = \#_k(A) \cdot \#_k(B), \quad \#_k(Z_n) = (k, n).$$

5.18. The operations **P**, **M**, **A** are not independent: show that two rows can be transposed by a sequence of (four) operations of types **M** and **A**.

5.19. Prove that the group $Z_m \times Z_n$, $m, n \in \mathbb{N}$, is cyclic if and only if $(m, n) = 1$.

5.20. Let G be group with a **balanced presentation** $(m = n)$, so that the relation matrix E is square. Prove that $|G : G'| = \pm \det E$.

6
Discrete Subgroups of the Euclidean Group

Since the Euclidean group $\mathbb{E}$ has continuously many elements:

$$|\mathbb{E}| = |\mathbb{R}| = c,$$

it has 2^c subsets, and each of these generates a subgroup. So, to avoid bewilderment, it is necessary to place some restriction on the class of subgroups to be investigated. One such restriction, and a very natural one, was foreshadowed at the end of Chapter 1. A subgroup G of $\mathbb{E}$ is said to be **discrete** if, for any point O in $\mathbb{R}^2$, every circle centre O contains only finitely many points of the orbit $OG = \{Og \mid g \in G\}$. This means that OG has no accumulation points, and it follows that around every point $O \in \mathbb{R}^2$ there is a circle (of positive radius) containing no point of OG other than O itself. A practical consequence is that, for any point $O \in \mathbb{R}^2$ not fixed by G, *there is an element in OG of minimal positive distance from O.*

The classification of the discrete subgroups of $\mathbb{E}$ will occupy this chapter and the next two, which together form the centrepiece of the book. The story begins with an indubitably classical result.

6.1 Leonardo's Theorem

In this section, we study subgroups $G \le \mathbb{E}$ satisfying a different but related condition, namely, that $G \cap \mathbb{T} = \{1\}$, where $\mathbb{T}$ is the translation subgroup of $\mathbb{E}$. Then G contains no glide reflection q, as $1 \ne q^2 \in \mathbb{T}$. By Theorem 4.3, G therefore consists entirely of rotations and reflections. If G contains no rotation (other than 1), it follows from Theorem 4.1 that it contains at most one reflection. In this case, G is either trivial or cyclic of order 2. Otherwise, G contains a rotation $s \ne 1$, $s = s(O, \theta)$ say. We claim that every element of G fixes O, and this is proved by contradiction as follows.

Suppose G contains an element g that moves O: $O' = Og \ne O$. Then it is not hard to show (Exercise 6.1) that $g^{-1}sg = s(O', \pm\theta)$. It then follows from the discussion in Chapter 4 (see Fig. 4.5(i)) that $s^{\mp 1}g^{-1}sg$ is a non-trivial translation. But such elements are not allowed in G, and we have proved our claim.

Theorem 6.1

If $G \le \mathbb{E}$ and $G \cap \mathbb{T} = \{1\}$, then there is a point O in $\mathbb{R}^2$ fixed by every element of G. □

Corollary 6.1 (Leonardo da Vinci)

Every finite subgroup of $\mathbb{E}$ is either cyclic or dihedral.

Proof

Let G be a finite subgroup of $\mathbb{E}$. Then, since any non-trivial translation has infinite order, we must have $G \cap \mathbb{T} = \{1\}$, and it follows from the theorem that G fixes a point, O say: $Og = O$ for all $g \in G$. The OP subgroup G^+ of G thus consists of rotations about O. If $G^+ = \{1\}$, we have $G \cong \{1\}$ or Z_2, as above. Otherwise, we proceed as follows.

Among the finitely many rotations in G, there is one, $s = s(O, \theta)$ say, through minimal positive angle θ. By a theorem of Archimedes, there is a least $m \in \mathbb{N}$ for which $m\theta \ge 2\pi$. The minimality of m ensures that $m\theta < 2\pi + \theta$, and the minimality of θ that $m\theta = 2\pi$. The minimality of θ also ensures that G contains no rotation $s' = s(O, \phi)$ with $(n-1)\theta < \phi < n\theta$ for any $n \in \mathbb{N}$ (consider $s's^{-(n-1)}$). Thus, every rotation in G is a power of s, that is, $G^+ = \langle s \mid s^m \rangle \cong Z_m$.

If G contains no reflection, then $G = G^+ \cong Z_m$, as required. Otherwise, let $r \in G$ be reflection in a line l, which necessarily passes through O. Then, as in the proof of the theorem, $r^{-1}sr = s(O, -\theta) = s^{-1}$, whence

$$G = \langle s, r \mid s^m = 1, r^2 = 1, r^{-1}sr = s^{-r} \rangle \cong D_{2m},$$

and this completes the proof.   □

6.2 A Trichotomy

Theorem 6.2

If G is a discrete subgroup of $\mathbb{E}$, then its translation subgroup $T = G \cap \mathbb{T}$ is either trivial, infinite cyclic, or free abelian of rank 2:

$$T \cong \mathbb{Z}^r, \quad r = 0, 1, \text{ or } 2.$$

Proof

Assume that T is non-trivial, and fix a point $O \in \mathbb{R}^2$. Now T, being a subgroup of the discrete G, is itself discrete, and so there is an $a \in T \setminus \{1\}$ with

$$d(O, Oa) \leq d(O, Ot) \; \forall t \in T \setminus \{1\}. \tag{6.1}$$

Then the line l through O and Oa contains all the points Oa^n, $n \in \mathbb{Z}$, of the orbit of O under the infinite cyclic subgroup $H = \langle a \rangle$ of T. If $H = T$ we are finished, so assume that $H < T$ and proceed as follows.

First we claim that for no $t \in T \setminus H$ is $Ot \in l$. For otherwise there are two possibilities: for some $t \in T \setminus H$ either

(a) $Ot = Oa^n$ for some $n \in \mathbb{Z}$, or

(b) Ot lies in the interior of a segment $[Oa^n, Oa^{n+1}]$.

In the first case, $a^n t^{-1}$ is a translation fixing O, whence $a^n t^{-1} = 1$. Then $t = a^n \in H$, contradicting the choice of t. In the second case,

$$0 < d(O, Ota^{-n}) = d(Oa^n, Ot) < d(Oa^n, Oa^{n+1}) = d(O, Oa),$$

contradicting the minimality of $d(O, Oa)$ in (6.1). This proves our claim that $Ot \notin l \; \forall t \in T \setminus H$.

Next we again invoke discreteness to pick a $b \in T \setminus H$ with

$$d(O, Ob) \leq d(O, Ot) \; \forall t \in T \setminus H. \tag{6.2}$$

Then the line l' through O and Ob contains all the points Ob^n, $n \in \mathbb{Z}$, of the orbit of O under the infinite cyclic subgroup $K = \langle b \rangle$ of T. Since $l \cap l' = \{O\}$

by the claim proved above, the lines $l'a^m$ and lb^n, $m, n \in \mathbb{Z}$, partition $\mathbb{R}^2$ into parallelograms whose vertices form the orbit of O under the subgroup $H \times K$ of T. Now $H \times K \cong \mathbb{Z}^2$, so if $H \times K = T$ we are finished. So assume that $H \times K < T$ and get a contradiction as follows.

Let $t \in T \setminus H \times K$. Then Ot is in or on one of the parallelograms in the above partition, say that with vertices

$$Oa^m b^n, \ Oa^{m+1} b^n, \ Oa^{m+1} b^{n+1}, \ Oa^m b^{n+1},$$

$m, n \in \mathbb{Z}$. Then $P = Ota^{-m}b^{-n}$ is in or on the parallelogram $OAQB$, where $A = Oa$, $B = Ob$, $Q = Oab$ (see Fig. 6.1). Since $ta^{-m}b^{-n}$ lies outside $H \times K$, P is not equal to A or B. In the case when $d(O, P) \leq d(Q, P)$ (as in the figure), an elementary property of triangles (Exercise 6.6) then implies that $d(O, P) < d(O, B)$, which contradicts the minimality of b in (6.2). When $d(O, P) \geq d(Q, P)$, a similar contradiction is obtained from the location of the point $Pa^{-1}b^{-1}$.

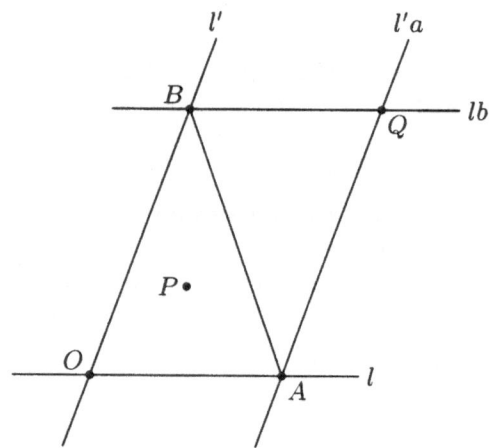

Figure 6.1   The parallelogram $OAQB$

We have thus shown that T is one of

$$\{1\} \cong Z^0, \ \langle a \rangle \cong Z, \ \langle a, b \rangle \cong Z^2,$$

as required.   □

To classify discrete subgroups G of $\mathbb{E}$ in the first of these three cases, namely, $G \cap T = \{1\}$, merely apply Theorem 6.1, Exercise 6.3 and Leonardo's Theorem in turn to get

$$G \text{ in Case } 1 \Rightarrow G \leq \mathrm{Sym}(O) \Rightarrow G \text{ finite} \Rightarrow G \cong Z_n \text{ or } D_{2n},$$

where $n \in \mathbb{N}$. We now turn to Case 2, when G is a discrete subgroup of $\mathbb{E}$ such that $G \cap T \cong Z$.

6.3 Friezes and Their Groups

The rest of this chapter is devoted to the study of discrete subgroups of $\mathbb{E}$ whose translation subgroup is infinite cyclic. They are the symmetry groups of certain infinite plane figures whose translational symmetries are just the iterates of some translation along an axis. Such figures are called **friezes** and the corresponding groups, **frieze groups**. For the remainder of this chapter, F will denote a frieze group and $T = \langle t \mid \rangle$ its translation subgroup.

Case 1: F contains no non-trivial rotation, so that $F^+ = T$. It may be that $F = T$, and we have our first type,

$$F_1 = \langle t \mid \rangle.$$

Otherwise, F contains an OR isometry r, which is either a reflection or a glide reflection. In either case, $F^- = Tr$, $|F : T| = 2$, and $T \triangleleft F$. Thus, $r^{-1}tr$ generates $r^{-1}Tr = T$, and this forces $r^{-1}tr = t^{\pm 1}$. In geometric terms (see the corollary to Theorem 2.4) this means that if $t = t(\mathbf{a})$, then $\mathbf{a}r = \pm\mathbf{a}$ and the axis of r is either parallel or perpendicular to the direction of $\mathbf{a}$. If r is a reflection, it may be of either kind, and we get two more types,

$$F_1^1 = \langle t, r \mid r^2 = 1, \ t^r = t \rangle,$$
$$F_1^2 = \langle t, r \mid r^2 = 1, \ t^r = t^{-1} \rangle.$$

Otherwise, r is a glide reflection and r^2 is a non-trivial translation, $r^2 = t^h$ for some $h \in \mathbb{Z}$, $h \neq 0$. Then r commutes with t^h and hence with t, and so, for all $k \in \mathbb{Z}$,

$$(rt^k)^2 = r^2 t^{2k} = t^{2k+h}. \tag{6.3}$$

At this stage we need to employ a trick, which consists of

(a) changing the generators (Tietze transformation),

(b) changing the relations accordingly,

(c) reallocating symbols.

In the case in hand, choose $k \in \mathbb{Z}$ such that the exponent $2k + h$ in (6.3) is either 0 or 1, and then

(a) retain t but replace r by $r' = rt^k$,

(b) observe that r' commutes with t,

(c) glibly redenote r' by r.

This trick will be used again later, where it will be referred to simply as "the trick".

As well as commuting with t, the new r has square equal to 1 or t. In the former case we get F_1^1 again, and in the latter we get the new type

$$F_1^3 = \langle t, r \mid r^2 = t, \ t^r = t \rangle.$$

This completes the classification of the rotationfree frieze groups into four geometrically different types. While two of these, F_1 and F_1^3, are isomorphic as groups, one contains OR isometries and the other does not.

Case 2: F contains a non-trivial rotation s. Since t^s is again a generator for T ($T^s = T$ as $T \vartriangleleft F$), we must have $t^s = t^{\pm 1}$. But this means that if $t = t(\mathbf{a})$, then $\mathbf{a}^s = \pm\mathbf{a}$, and since $s \neq 1$ we must take the minus sign. Thus s is a rotation through π and so has order 2. Any other rotation $s' \in F$ likewise has order 2, whence $s's$ is a translation. Then $s' \in Ts$ and it follows that $|F^+ : T| = 2$. It may be that $F = F^+$, and we get the type

$$F_2 = \langle t, s \mid s^2 = 1, \ t^s = t^{-1} \rangle.$$

Otherwise, $F = F^+ \cup F^+ r$, where r is a reflection or glide reflection. Let l be the axis of t through the centre O of s, and l' the axis of r. As in Case 1, $t^r = t^{\pm 1}$ and we deduce that l' is either parallel or perpendicular to l. In the latter case, $t^{sr} = t$ and sr is an OR isometry with axis parallel to l. Taking sr instead of r by the trick, we can thus assume that l' is parallel to l. We claim that $l = l'$.

Assume that $l' \neq l$ and look for a contradiction. The assumption implies that $Or \notin l$, and the composite

$$ss^r = s(O, \pi)s(Or, \pi)$$

is a non-trivial translation t' along the line m through O and Or. But since $l \cap m = \{O\}$, l and m are not parallel and $t' \notin T$. This contradiction proves the claim.

If r is a reflection, then so is sr, with axis m perpendicular to l at 0, and we get the type

$$F_2^1 = \langle t, s, r \mid s^2 = 1, t^s = t^{-1}, r^2 = 1, t^r = t, (sr)^2 = 1 \rangle.$$

Replacing the relation $(sr)^2 = 1$ by $s^r = s$, which is equivalent, the resulting presentation is of the form (5.10) and exhibits F_2^1 as a semidirect product of the form $(Z_2 \times Z_2) \times_\alpha Z$.

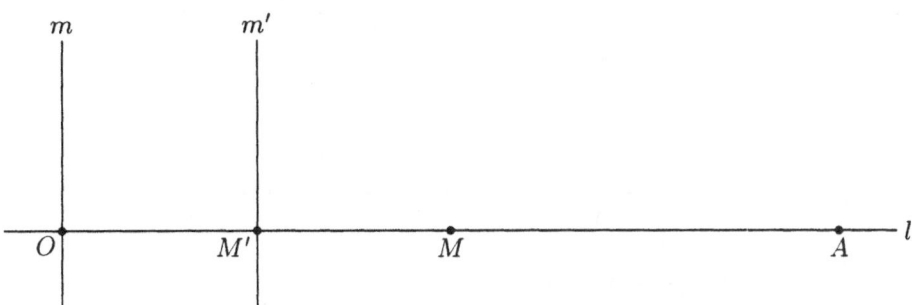

Figure 6.2   Construction of m'

But if r is a glide reflection, we can assume by the trick that $r^2 = t$. Now recall that l is the axis of r. Let A be the point Ot, M the midpoint of OA, and M' the midpoint of OM as in Fig. 6.2. Moreover let m, m' be the lines perpendicular to l through O, M' respectively. Then we have

$$s = r(m)r(l), \quad r = r(l)r(m)r(m').$$

It follows that $sr = r(m')$, whence $(sr)^2 = 1$ and we get our final type

$$F_2^2 = \langle t, s, r \mid s^2 = 1, t^s = t^{-1}, r^2 = t, t^r = t, (sr)^2 = 1 \rangle.$$

6.4 The Classification

Theorem 6.3

There are exactly seven geometrically different types of frieze, as illustrated below along with presentations for their groups.   □

In each of the friezes in Fig. 6.3, the axis l is represented by a horizontal line. The images of the configuration are understood to continue indefinitely along l in both directions.

EXERCISES

6.1. Let $s, g \in \mathbb{E}$, where s is the rotation $s(O, \theta)$ and g is arbitrary. Using considerations in Chapter 2 or otherwise, prove that $g^{-1}sg = s(Og, \pm\theta)$, where we take $+$ or $-$ according as g is OP or OR.

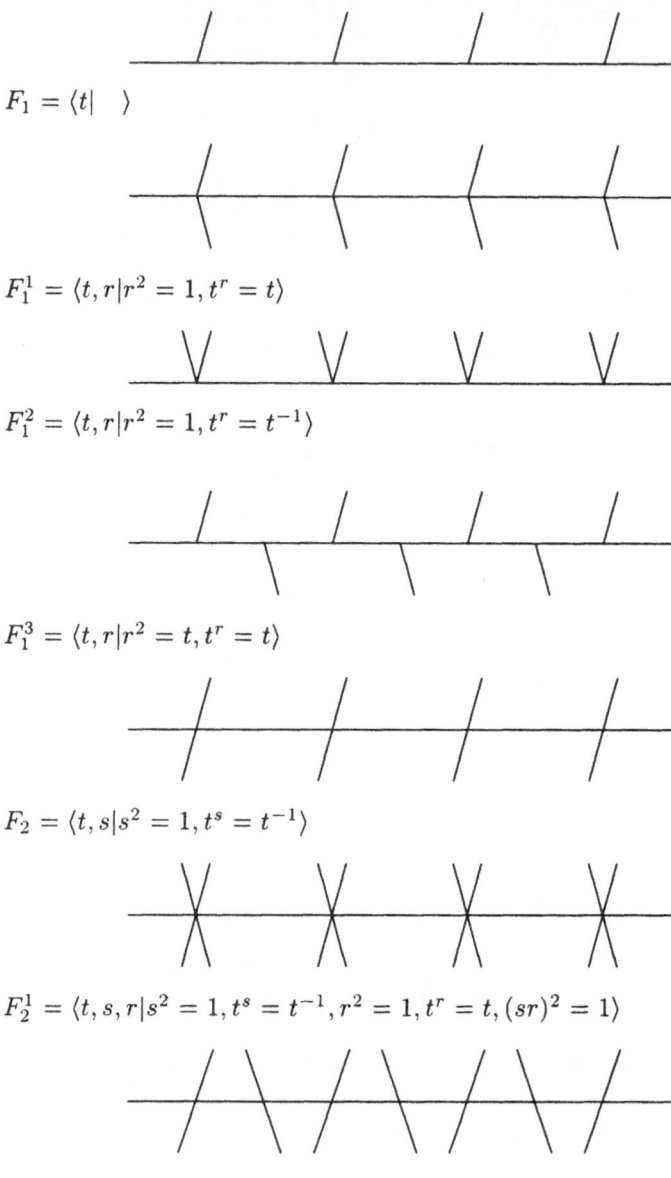

$$F_1 = \langle t| \quad \rangle$$

$$F_1^1 = \langle t, r | r^2 = 1, t^r = t \rangle$$

$$F_1^2 = \langle t, r | r^2 = 1, t^r = t^{-1} \rangle$$

$$F_1^3 = \langle t, r | r^2 = t, t^r = t \rangle$$

$$F_2 = \langle t, s | s^2 = 1, t^s = t^{-1} \rangle$$

$$F_2^1 = \langle t, s, r | s^2 = 1, t^s = t^{-1}, r^2 = 1, t^r = t, (sr)^2 = 1 \rangle$$

$$F_2^2 = \langle t, s, r | s^2 = 1, t^s = t^{-1}, r^2 = t, t^r = t, (sr)^2 = 1 \rangle$$

Figure 6.3   The seven friezes and their groups

6.2. Prove that if G is a discrete subgroup of $\mathbb{E}$ and H is a subgroup of G, then H is discrete.

6.3. Prove that a discrete subgroup of $\mathbb{E}$ fixes a point if and only if it is finite.

6.4. Define and classify the discrete subgroups of $\mathrm{Isom}(\mathbb{R}^1)$.

6.5. Make a conjecture about the translation subgroup of a discrete subgroup of $\mathrm{Isom}(\mathbb{R}^n)$, $n \geq 3$, but do not on any account try to prove it.

6.6. Prove that a point in or on a plane triangle of maximal distance from one of the vertices can only be one (or both) of the other two.

6.7. For each of the finite subgroups $G \cong Z_n$ or D_{2n} of $\mathbb{E}$, draw a plane figure C such that $G = \mathrm{Sym}(C)$.

6.8. Prove that if G is any subgroup of $\mathbb{E}$, then $|G : G^+| = 1$ or 2.

6.9. Show that for any subgroup G of $\mathbb{E}$, the requirement that $G \cap \mathbb{T} \cong Z$ forces G to be discrete.

6.10. For each frieze group F, describe the elements of F in the normal form of Chapter 2. Identify the elements of finite order in each case.

6.11. Recall from Chapter 3 that for any group G the centre $Z(G)$ is the subgroup of G consisting of those $z \in G$ such that $gz = zg$ for all $g \in G$. Describe $Z(F)$ for each frieze group F.

6.12. For each frieze group F, throw enough commutators into its presentation to be able to describe F/F', then identify F'.

6.13. Guided by the last three exercises, show that the seven frieze groups fall into exactly four isomorphism classes.

6.14. Say which of the frieze groups contain glide reflections.

6.15. Let $L \subseteq \mathbb{R}^2$ be the infinite ladder shown in Fig. 6.4. Write down enough symmetries of L to be able to identify $\mathrm{Sym}(L)$ among the seven frieze groups.

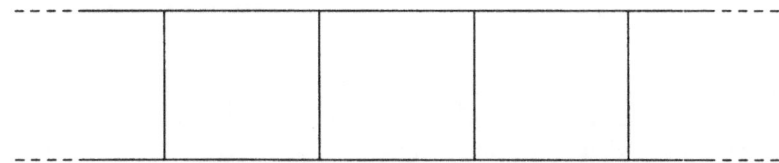

Figure 6.4   The infinite ladder

6.16. This is a five-part exercise.

   (a) Go to your local airport and get on a plane.

   (b) Get off at Madrid and take a bus to Granada.

   (c) Find the Alhambra Palace and go inside.

   (d) Take pictures of the many friezes to be seen there.

   (e) Compare your results with the diagrams in Fig. 6.3.

7
Plane Crystallographic Groups: OP Case

A discrete subgroup G of $\mathbb{E}$ with $T = G \cap \mathbb{T} \cong Z^2$ is called a **plane crystallographic group**. Such groups will be classified in this chapter and the next, thereby completing the description of all discrete subgroups of $\mathbb{E}$ in accordance with Theorem 6.2. As in the proof of that theorem, we put $T = \langle a, b \rangle$, where a, b are translations in $T \setminus \{1\}$, $T \setminus \langle a \rangle$, respectively, through minimal distance. Then G may be computed as a twofold extension of T, first to G^+ then to G. The first of these steps will occupy this chapter, so we assume for its duration that $G = G^+$.

7.1 The Crystallographic Restriction

The assumption that $G = G^+$ means that every element of $G \setminus T$ is a rotation. Of course $G \setminus T$ may be empty, in which case we get our first type,

$$G_1 = \langle a, b \mid ab = ba \rangle.$$

So assume from now on that $T \neq G$, that is, G contains non-trivial rotations. The effect on rotations of the assumption of discreteness is made clear in the following lemma.

Lemma 7.1

If s is a rotation lying in a discrete subgroup of $\mathbb{E}$, then

(i) *s has finite order, say $n \in \mathbb{N}$, and*

(ii) *some power of s is equal to $s(O, 2\pi/n)$, where O is the centre of s.*

Proof

(i) Suppose for a contradiction that s has infinite order, so that the positive powers s^n, $n \in \mathbb{N}$, are all distinct. Then any point $P \neq O$ has infinitely many images Ps^n, $n \in \mathbb{N}$, on the circle C centre O radius $d(O, P)$. But then the circle centre P radius $2d(O, P)$, which contains C, must contain infinitely many images of its centre P under powers of s, contradicting discreteness.

(ii) Let $s = s(O, \theta)$ and put $s_1 = s(O, 2\pi/n)$. Since $1 = s^n = s(O, n\theta)$, it follows that $n\theta = 2k\pi$, $k \in \mathbb{Z}$. Then $\theta = 2k\pi/n$, whence $s = s_1^k$ and we have proved the wrong thing. But wait. The subgroups $\langle s \rangle$ and $\langle s_1 \rangle$ both have finite order n, and we have just shown that $\langle s \rangle \leq \langle s_1 \rangle$. So the reverse inclusion, which is what we want, follows from the pigeon-hole principle. □

It follows from part (i) that any rotation in $G \setminus T$ generates a finite subgroup $H \leq G$. Part (ii) then implies that H contains a rotation s through minimal positive angle $2\pi/n$, where $n = |H|$. Our next task is to prove that this n can only be 2, 3, 4 or 6. This limitation, especially the exclusion of 5, is known as the **crystallographic restriction**.

Lemma 7.2

The order n of a non-trivial rotation in G is 2, 3, 4 or 6.

Proof

Any such rotation is a power of $s = s(O, 2\pi/n)$, where O is the centre and $n \in \mathbb{N}$, $n \geq 2$. Set $A = Oa$, where a is a minimal translation, and consider the circle centre O through A. This image of A under powers of s all lie on this circle, as illustrated in the cases $n = 5$, $n \geq 7$ in Fig. 7.1. We seek a contradiction in each of these cases in turn.

Assume first that $n \geq 7$ and consider the translation $t = a^{-1}s^{-1}as$. We compute

$$At = Oaa^{-1}s^{-1}as = Os^{-1}as = Oas = As,$$

so that

$$d(O, Ot) = d(A, At) = d(A, As) < d(O, A) = d(O, Oa),$$

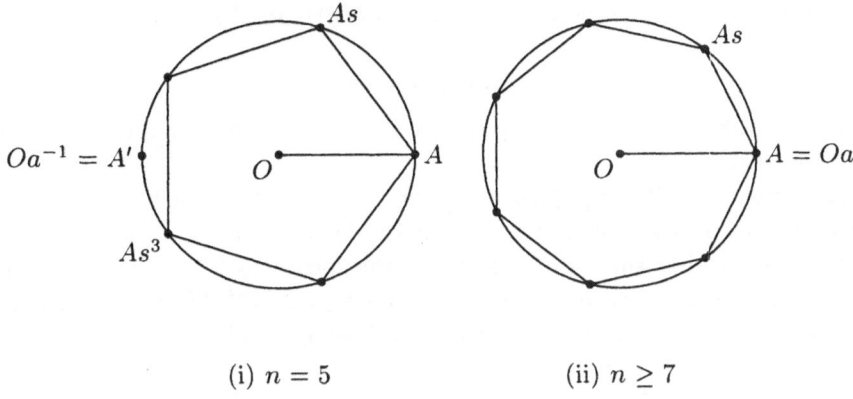

(i) $n = 5$ (ii) $n \geq 7$

Figure 7.1   The crystallographic restriction

since the side of a regular n-gon is less than its radius when $n \geq 7$. This contradicts the minimality of a and rules out case (ii).

Now assume that $n = 5$ and consider the translation $t = as^{-3}as^3$. Setting $Oa^{-1} = A'$, we compute

$$A't = Oa^{-1}as^{-3}as^3 = Os^{-3}as^3 = Oas^3 = As^3,$$

so that

$$d(O, Ot) = d(A', A't) = d(A', As^3) < d(O, A) = d(O, Oa),$$

by the above property of a regular decagon. This again contradicts the minimality of a, and case (i) is also excluded.   □

7.2 The Parameter n

Let n denote the maximal order of a rotation in G. Then, by Lemma 7.1(ii), we may assume that G contains a rotation s through $2\pi/n$, say $s = s(O, 2\pi/n)$. The powers of s all have orders dividing n, and we claim that these are the only orders possible for a rotation in G.

For suppose that G contains a rotation of order m, $m \nmid n$. Then, again by Lemma 7.1(ii), G contains a rotation s' through $2\pi/m$. Since $m \nmid n$, the highest common factor h of m and n is less than m. By Euclid's algorithm, we can find integers c, d such that

$$cm + dn = h < m.$$

By the discussion in Section 4.3 (see Fig. 4.5(ii)), $s^c s'^d$ is a rotation through the angle

$$2\pi c/n + 2\pi d/m = 2\pi h/mn,$$

which has order $mn/h > n$. This contradiction to the maximality of n establishes the claim.

Now fix $s = s(O, 2\pi/n)$ as above and let $g \in G$ be arbitrary. Then g is either a translation or a rotation. In the latter case, $|g| = d$ with $d \mid n$, and g is a rotation through an angle of the form $2\pi k/d$, $k \in \mathbb{Z}$. As in the previous paragraph (see Fig. 4.5(i)), it follows that $s^{-kn/d}g$ is a translation. Then $g \in s^{kn/d}T$ and we have proved the following result.

Lemma 7.3

The translation subgroup T of G has a complement $S = \langle s \rangle \cong Z_n$ in G, where n is the maximal order of a rotation in G.   □

Combining this with Lemma 7.2, we see that G is a semidirect product of the form $Z_n \times_\alpha Z^2$, where $n = 2, 3, 4$ or 6. It remains to determine the action α, and this is done by judicious choice of the generator b relative to a and s.

7.3 The Choice of b

The case $n = 2$ presents no difficulty: s is a rotation through π, so that conjugation by s must send any translation to its inverse. In particular,

$$s^{-1}as = a^{-1}, \qquad s^{-1}bs = b^{-1},$$

and these equations determine the action α of s on T. We have our second type,

$$G_2 = \langle a, b, s \mid ab = ba, s^2 = 1, a^s = a^{-1}, b^s = b^{-1} \rangle.$$

Note that the choice of $b \in T \setminus \langle a \rangle$ here, as in the case of G_1, is subject only to the minimality requirement. This is not so in the remaining cases, which are coming shortly. What is so, in all (five) cases, is that the group is determined by the parameter n.

The cases $n = 3, 4, 6$ are illustrated in Fig. 7.2, where $s = s(O, 2\pi/n)$. In each case the figure depicts the images under powers of s of the point $A = Oa$ on the circle centre O radius $d(O, A)$. Note that each of these points is the image of O under a translation: for example,

$$As = Oas = Os^{-1}as,$$

the image of O under $s^{-1}as$. Since all these translations are through the same (minimal) distance as a, any of these points other than $Oa^{\pm 1}$ is a candidate for Ob. We choose b in each case as follows and put $B = Ob$.

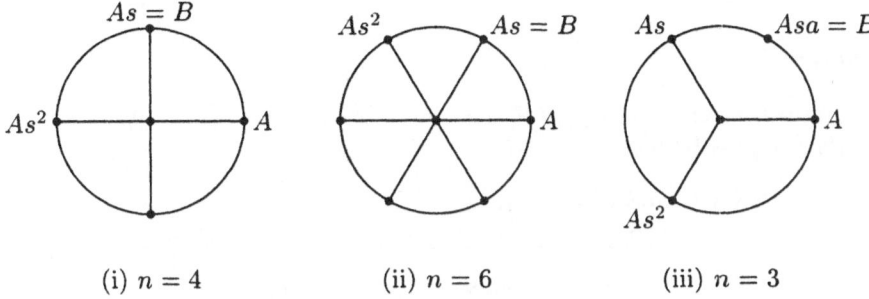

(i) $n = 4$ (ii) $n = 6$ (iii) $n = 3$

Figure 7.2   The choice of b

(i) $n = 4$: take b to be the translation $s^{-1}as$ sending O to As. Then $s^{-1}bs$ is the translation sending O to

$$Os^{-1}bs = Obs = As^2 = Oa^{-1}.$$

Thus the action of S on T is given by

$$s^{-1}as = b, \quad s^{-1}bs = a^{-1}.$$

(ii) $n = 6$: again let $b = s^{-1}as$, which sends O to As. Since As^2 is the image under a^{-1} of $As = Ob$, $s^{-1}bs$ sends O to

$$Os^{-1}bs = Obs = As^2 = Oba^{-1} = Oa^{-1}b,$$

as a and b commute. Then the action of S on T is given by

$$s^{-1}as = b, \quad s^{-1}bs = a^{-1}b.$$

(iii) $n = 3$: for consistency with case (ii), we take Ob to be the point B on the circle diametrically opposite As^2. Then

$$Ob^{-1} = As^2 = As^{-1} = Osas^{-1},$$

and, since $As = Ba^{-1}$,

$$Os^{-1}as = As = Oba^{-1} = Oa^{-1}b.$$

Thus the action of S on T is given by

$$s^{-1}as = a^{-1}b, \quad s^{-1}bs = a^{-1}.$$

7.4 Conclusion

We have shown that each of the five cases $n = 1, 2, 3, 4, 6$ leads to a single group G_n, with presentation as in the following theorem.

Theorem 7.1

If G is a discrete subgroup of $\mathbb{E}$ with $T \cong Z^2$ and $G^+ = G$, then there are exactly five possibilities:

$$G_1 = \langle a, b \mid ab = ba \rangle,$$
$$G_2 = \langle a, b, s \mid ab = ba, s^2 = 1, a^s = a^{-1}, b^s = b^{-1} \rangle,$$
$$G_3 = \langle a, b, s \mid ab = ba, s^3 = 1, a^s = a^{-1}b, b^s = a^{-1} \rangle,$$
$$G_4 = \langle a, b, s \mid ab = ba, s^4 = 1, a^s = b, b^s = a^{-1} \rangle,$$
$$G_6 = \langle a, b, s \mid ab = ba, s^6 = 1, a^s = b, b^s = a^{-1}b \rangle.$$

$\square$

As in the case of the frieze groups, each of these groups is the symmetry group of a plane figure. Such figures are called **tessellations**, or tilings, of $\mathbb{R}^2$. The choice of suitable tessellations affords considerable scope for artistic license, an indulgence we postpone until Chapter 9.

EXERCISES

7.1. Show that for any subgroup G of $\mathbb{E}$, the requirement that $G \cap T \cong Z^2$ forces G to be discrete.

7.2. For each $n = 2, 3, 4, 6$, write down a 2×2 matrix over Z that describes the action of s on T in G_n.

7.3. For each n describe the elements of G_n in normal form. Identify the elements of finite order in each case.

7.4. For each i, j, prove that G_i is isomorphic to a subgroup of G_j if and only if i is a divisor of j.

7.5. Prove that the five groups listed in Theorem 7.1 are pairwise non-isomorphic.

7.6. For each n identify the centre of G_n.

7.7. For each n describe the derived factor group G_n/G'_n, then identify G'_n.

7.8. Show that all but one of the G_n can be generated by two elements and write down a 2-generator presentation for each of these four groups.

7.9. Let G be an arbitrary (not necessarily OP) discrete subgroup of $\mathbb{E}$ with translation subgroup T. Prove that $|G : T| \leq 12$.

7.10. Let a, b be translations through minimal distances in directions that are neither parallel nor perpendicular. Use a, b to describe a tessellation of $\mathbb{R}^2$ whose symmetry group is G_2.

Plane Crystallographic Groups: OR Case

In this chapter G is a discrete subgroup of $\mathbb{E}$ with translation subgroup $T \cong Z^2$ and OP subgroup G^+ of index 2. Then $G = G^+ \cup G^+r$, where r is OR and G^+ is one of the groups G_n, $n = 1, 2, 3, 4, 6$, derived in the previous chapter. Thus G is determined by the values of n, r^2 and of the conjugates a^r, b^r, s^r. As usual the treatment is case by case, greatly simplified by the following useful dichotomy.

8.1 A Useful Dichotomy

We focus on the action of r on T, and limit the number of possibilities for the pair (a^r, b^r) to two. This is done by the trick described in Chapter 6: choose new generators of T and, if necessary, of S and reallocate the symbols a, b, s accordingly. The meaning of these three symbols will thus change in the course of the next two pages.

Lemma 8.1

For a fixed $r \in G \setminus G^+$, exactly one of the following two conditions holds:
   T is generated by elements a, b such that

$$a^r = a, \quad b^r = b^{-1},\tag{8.1}$$

T is generated by elements a, b such that

$$a^r = b, \quad b^r = a. \tag{8.2}$$

Proof

We show first that (8.1) and (8.2) are incompatible. Suppose, for a contradiction that (8.2) holds for generators a, b and (8.1) for generators $a' = a^i b^j$, $b' = a^k b^l$, where $i, j, k, l \in \mathbb{Z}$. Then

$$a^i b^j = a' = a'^r = a^j b^i,$$

whence $i = j$. Similarly,

$$a^k b^l = b' = (b'^r)^{-1} = a^{-l} b^{-k},$$

whence $k = -l$. Thus T is generated by the pair

$$a' = (ab)^i, \quad b' = (ab^{-1})^k$$

and hence by ab, ab^{-1}. Then for some $m, n \in \mathbb{Z}$,

$$a = (ab)^m (ab^{-1})^n = a^{m+n} b^{m-n}.$$

The fact that a, b are free generators now implies that $m + n = 1$, $m - n = 0$, which yields the contradiction $2m = 1$. So (8.1) and (8.2) cannot both hold for the same r in the same group.

Now let $a, b \in T$ be translations through minimal distance in $T \setminus \{1\}$, $T \setminus \langle a \rangle$ respectively, as above. Fix a point $O \in \mathbb{R}^2$ and put $Oa = A$, $Ob = B$. Replacing a, b by $a^{\pm 1}, b^{\pm 1}$ as necessary, we can assume that the angle $\theta = A\widehat{O}B$ lies between 0 and $\pi/2$. By minimality of b, B is not closer to A than to O, and so lies on the same side of the perpendicular bisector l of O, A as O (see Fig. 8.1(i)). In fact, by minimality of a, B cannot lie in the interior of the equilateral triangle OAC, and so $\pi/3 \le \theta \le \pi/2$.

We now distinguish three cases according as the axes of a and r are parallel (so $a^r = a$), perpendicular (so $a^r = a^{-1}$) or neither (so $a^r \ne a^{\pm 1}$).

(i) Let $a^r = a$. If $\theta = \pi/2$, then $b^r = b^{-1}$ and (8.1) holds with the original (a, b). But if $\theta < \pi/2$ (see Fig. 8.1(ii)), then $d(Ob^r, Oab^{-1}) < d(O, Oa)$, and the minimality of a forces $b^r = ab^{-1}$. Then

$$(ab^{-1})^r = a^r b^{-r} = a(ab^{-1})^{-1} = b,$$

and (8.2) holds for the pair (b, ab^{-1}).

(ii) Let $a^r = a^{-1}$. Then we see from Fig. 8.1(iii) that $d(Ob^r, Ob) \le d(O, Oa)$, and the minimality of a means that either $b^r = b$ or $b^r a = b$. This means that (8.1) holds for (b, a) or (8.2) holds for $(b, a^{-1} b)$ respectively.

(iii) Let $a^r \ne a^{\pm 1}$, so that $a^r \notin \langle a \rangle$ and we can take $b = a^r$. Then $b^r = a^{r^2} = a$, and (8.2) holds for (a, b). $\qquad\square$

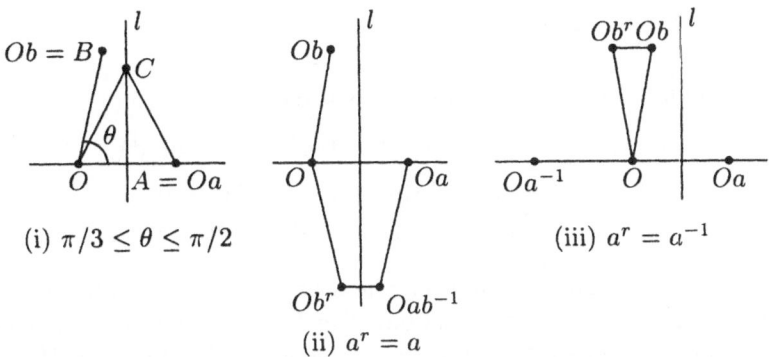

(i) $\pi/3 \le \theta \le \pi/2$

(iii) $a^r = a^{-1}$

(ii) $a^r = a$

Figure 8.1 Action of r on a, b

We must now examine the effect of these changes on s. First note that the substitution of a^{-1} and/or b^{-1} is only necessary when $n = 1$ or 2, since we already have $\theta = \pi/2$ when $n = 4$ and $\theta = \pi/3$ when $n = 3$ or 6. But s is absent when $n = 1$ and inverts every translation when $n = 2$, so these substitutions have no effect. Otherwise there are five changes possible, two each in cases (i) and (ii) and one in case (iii). We summarize the results in the form of a table.

Table 8.1 The effect on s

New a, b	θ	n	Effect on s	Case	Axis of r
a, b	$\pi/2$	$\neq 3, 6$	None	(8.1)	$\parallel$ axis of a
b, ab^{-1}	$< \pi/2$	$\neq 4$	Replace by s^{-1}	(8.2)	$\parallel$ axis of a
b, a	$\pi/2$	$\neq 3, 6$	Replace by s^{-1}	(8.1)	$\perp$ axis of a
$b, a^{-1}b$	$< \pi/2$	$\neq 4$	None	(8.2)	$\perp$ axis of a
a, b	Any	Any	None	(8.2)	$\parallel$ axis of ab

We proceed to examine the five cases $n = 1, 2, 3, 4, 6$ in turn, bearing in mind the following points:

(a) the relations obtained above hold for every $G^+ = G_n$,

(b) the relations (8.1) or (8.2) of Lemma 8.1 hold in every G except one,

(c) the value of r^2, when not 1, is given by the last column of Table 8.1,

(d) given (a), (b) and (c), G is determined by the value of s^r, or of $(sr)^2$.

These constraints correctly suggest that the number of such groups is finite. To find out exactly how many there are, read on.

8.2 The Case $n = 1$

This is the simplest case, and contains three types:

$$G_1^1 = \langle a, b, r \mid ab = ba, r^2 = 1, a^r = a, b^r = b^{-1} \rangle,$$
$$G_1^2 = \langle a, b, r \mid ab = ba, r^2 = 1, a^r = b, b^r = a \rangle,$$
$$G_1^3 = \langle a, b, r \mid ab = ba, r^2 = a, a^r = a, b^r = b^{-1} \rangle.$$

Here $G^+ = T$ is abelian, and any element $r \in G \setminus G^+$ induces the same automorphism of T. If r is a reflection, then Lemma 8.1 yields types G_1^1 and G_1^2. So assume that $G \setminus G^+$ contains no reflection and put $r^2 = a^k b^l$, $k, l \in \mathbb{Z}$. Then under condition (8.1) we have

$$a^k b^l = (a^k b^l)^r = a^k b^{-l},$$

whence $l = 0$. So $r^2 = a^k$ and, by the trick, we can take $k = 0$ or 1. But $k \neq 0$ by assumption, and we get the type G_1^3. On the other hand, condition (8.2) implies that

$$a^k b^l = (a^k b^l)^r = a^l b^k,$$

whence $k = l$. So $r^2 = (ab)^k$ and again we can take $k = 0$ or 1. We must have $k = 1$ as above, but then

$$(ra^{-1})^2 = r^2 \cdot r^{-1} a^{-1} r \cdot a^{-1} = ab \cdot b^{-1} \cdot a^{-1} = 1$$

contrary to assumption, and there is no group here.

8.3 The Case $n = 2$

This is the most populous case and contains four types, with generators a, b, s, r and relations

$$ab = ba, \quad s^2 = 1, \quad a^s = a^{-1}, \quad b^s = b^{-1}$$

plus, in

$$
\begin{aligned}
G_2^1, \qquad & r^2 = 1, \ a^r = a, \ b^r = b^{-1}, \ (sr)^2 = 1 \rangle, \\
G_2^2, \qquad & r^2 = 1, \ a^r = a, \ b^r = b^{-1}, \ (sr)^2 = b \rangle, \\
G_2^3, \qquad & r^2 = 1, \ a^r = b, \ b^r = a, \ (sr)^2 = 1 \rangle, \\
G_2^4, \qquad & r^2 = a, \ a^r = a, \ b^r = b^{-1}, \ (sr)^2 = b \rangle.
\end{aligned}
$$

Suppose first that G contains a reflection r for which condition (8.1) holds. Then $r^2 = 1$ and the axes of a and r are parallel. Since $b^{sr} = (b^{-1})^r = b$, the axes of sr and b are parallel, and so $(sr)^2 = b^k$, $k \in \mathbb{Z}$. As usual, we can take $k = 0$ or 1, giving types G_1^1 and G_1^2.

Suppose next that G contains a reflection r acting on a, b as in condition (8.2). Then r and ab have parallel axes and so do sr and $a^{-1}b$, since $(a^{-1}b)^{sr} = (ab^{-1})^r = a^{-1}b$. Hence, as usual, $(sr)^2 = 1$ or $a^{-1}b$. In the first case we have type G_2^3, and in the second we reallocate the symbol s to sb. The relations $s^2 = 1$, $a^s = a^{-1}$, $b^s = b^{-1}$ continue to hold, while $(sr)^2$ now denotes the old

$$(sbr)^2 = (sra)^2 = (sr)^2 a^{sr} a = a^{-1}b(a^{-1})^r a = a^{-1}bb^{-1}a = 1,$$

and we get G_2^3 again.

Suppose finally that G contains no reflection. Then under condition (8.1) we can take $r^2 = a$ and similarly $(sr)^2 = b$. It is an exercise to check that this type, which is G_2^4, contains no reflection. Under condition (8.2) we may assume as before that $r^2 = ab$ and $(sr)^2 = a^{-1}b$. But then

$$(sra)^2 = (sr)^2 a^{sr} a = a^{-1}bb^{-1}a = 1$$

contrary to assumption, and there is no group here.

8.4 The Case $n = 4$

This straightforward case contains two types, with generators a, b, s, r and relations

$$ab = ba, \quad s^4 = 1, \quad a^s = b, \quad b^s = a^{-1}$$

plus, in

$$G_4^1, \qquad r^2 = 1, \ a^r = a, \ b^r = b^{-1}, \ (sr)^2 = 1,$$
$$G_4^2, \qquad r^2 = a, \ a^r = a, \ b^r = b^{-1}, \ (sr)^2 = 1.$$

It follows from the relations defining G_4 that if $r \in G \setminus G^+$ acts on a, b as in (8.2), then sr acts on a, b as in (8.1). So we may ignore condition (8.2) and assume that $a^r = a$, $b^r = b^{-1}$. Then, as usual, $r^2 = 1$ or a and $(sr)^2 = 1$ or $a^{-1}b$, since

$$(a^{-1}b)^{sr} = (b^{-1}a^{-1})^r = ba^{-1} = a^{-1}b.$$

When $(sr)^2 = 1$ we get types G_4^1, G_4^2. If $(sr)^2 = a^{-1}b$, reallocate the symbol s to sa. This preserves the relations defining G_4, while $(sr)^2$ now denotes the old

$$(sar)^2 = (sra)^2 = (sr)^2 a^{sr} a = a^{-1}bb^{-1}a = 1,$$

giving nothing new.

8.5 The Case $n = 3$

This rather tricky case again contains two types, with generators a, b, s, r and relations

$$ab = ba, \quad s^3 = 1, \quad a^s = a^{-1}b, \quad b^s = a^{-1}$$

plus, in

$$
\begin{aligned}
G_3^1, \qquad & r^2 = 1, \ a^r = a, \ b^r = ab^{-1}, \ (sr)^2 = 1, \\
G_3^2, \qquad & r^2 = 1, \ a^r = b, \ b^r = a, \ (sr)^2 = 1.
\end{aligned}
$$

According to the description of G_3 in the previous chapter, $s = s(O, 2\pi/3)$ and the images of O under the translations $a^{\pm 1}$, $b^{\pm 1}$, $(a^{-1}b)^{\pm 1}$ are as shown in Fig. 8.2.

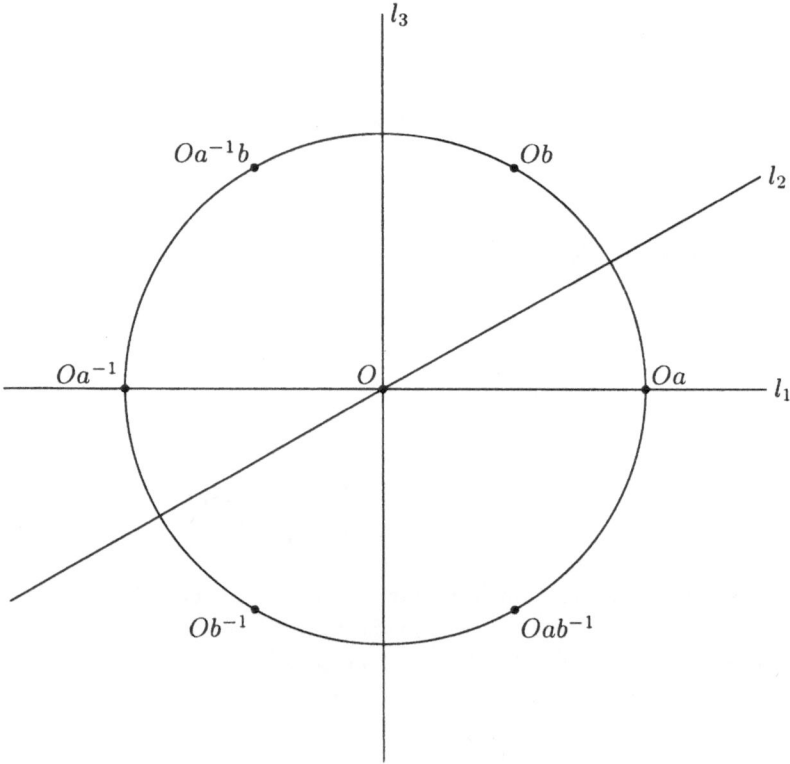

Figure 8.2   Translations through minimal distance in G_3

If $r \in G \setminus G^+$ has axis parallel to that of a, then we must have $b^r = ab^{-1} \neq b^{-1}$, and so condition (8.1) cannot hold when $n = 3$. So we may assume condition (8.2) for some pair of generators of T. Referring to Table 8.1, there are just three possibilities for such a pair,

$$(1)\ b, ab^{-1}, \quad (2)\ a, b, \quad (3)\ b, a^{-1}b$$

in terms of the original minimal generators a, b. The corresponding $r \in G \setminus G^+$ thus has axis parallel to one of the lines l_1, l_2, l_3 in Fig. 8.2. But since $s^2 = r(l_3)r(l_2)$, the effect on T of conjugation by rs^2 in case (3) is the same as that by r in case (2), and so it suffices to consider cases (1) and (2). Note that in the latter, r acts on T as in type G_3^2, while in the former

$$b^r = ab^{-1}, \quad a^r = (ab^{-1} \cdot b)^r = b \cdot ab^{-1} = a,$$

as in type G_3^1. Time for another useful lemma.

Lemma 8.2

Suppose that G contains a rotation s through $2\pi/3$ and a glide reflection r with axis l. Then G contains a reflection $r' = rt'$, $t' \in T$, with axis parallel to l.

Proof

Suppose that $r = r_0 t_0$, where r_0 and t_0 are reflection in and a translation along l respectively. (Note that r_0, t_0 may not be in G.) Since r is a glide reflection, t_0 is a translation through distance $d > 0$. Let l_1 and l_2 be the two lines parallel to and distant $\sqrt{3}d/2$ from l (see Fig. 8.3), and let P be an arbitrary point on l_1. Then $r^2 = t$ is translation along l through distance $2d = d(P, Pr)$, and t^s is translation along ls through the same distance. Since ls is parallel to the line through Pr, P, t^s sends Pr to P. It follows that rt^s fixes an arbitrary point P of l_1 and so is equal to $r(l_1)$. Taking $t' = t^s$ gives the required r'.   □

Now back to G_3. Modifying r in accordance with Lemma 8.2 if necessary, the effect on T of conjugation by r remains unchanged, and now $r^2 = 1$. To modify sr in a similar way, reallocate the symbol s to $s(sr)^2 = s^2 s^r$. Since $(sr)^2$ is a translation, the effect of conjugation on T is not changed, and since the total angle of rotation of the new s is still $2\pi/3$, the relation $s^3 = 1$ is also preserved. Finally, $(sr)^2$ now denotes the old

$$(s(sr)^2 r)^2 = (s^2 rs)^2 = (r^s)^2 = 1$$

as $r^2 = 1$. The relations $r^2 = (sr)^2 = 1$, with the action of r on T given by (8.1) and (8.2), give types G_3^1 and G_3^2.

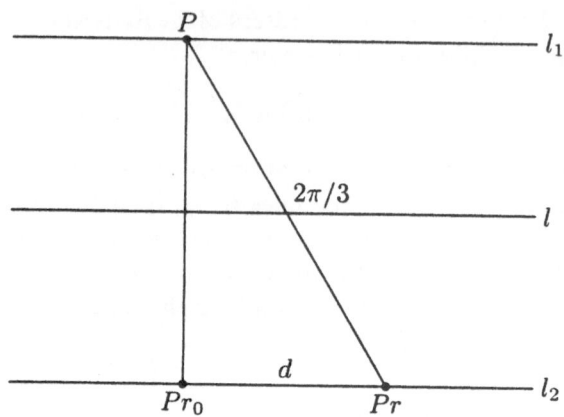

Figure 8.3   Construction of r'

8.6 The Case $n = 6$

In this final case there is just one type, with generators a, b, s, r and relations

$$ab = ba, \quad s^6 = 1, \quad a^s = b, \quad b^s = a^{-1}b$$

plus, in

$$G_6^1, \qquad r^2 = 1, \quad a^r = b, \quad b^r = a, \quad (sr)^2 = 1.$$

As in the previous case, the images of O under $a^{\pm 1}$, $b^{\pm 1}$, $(a^{-1}b)^{\pm 1}$ are as shown in Fig. 8.2, where O is now the centre of $s = s(O, \pi/3)$. Again condition (8.1) fails, any $r \in G \setminus G^+$ has axis parallel to l_1, l_2 or l_3, and b^r is ab^{-1}, a or $a^{-1}b$ respectively. As before, replacing r by rs^4 reduces (3) to (2). Moreover, we now have $s = r(l_1)r(l_2)$, so that, replacing r by rs reduces (1) to (2). Hence it suffices to consider case (2), where the action of r transposes a and b.

Since Lemma 8.2 continues to hold when $n = 6$, we may assume that $r^2 = 1$. To modify sr in a similar way, note that the relations defining G_6 are preserved when we reallocate the symbol s to st^r for any $t \in T$. Then sr now denotes the old srt, and it remains to choose t as in Lemma 8.2 to ensure that the new sr is a reflection. The final result is the group G_6^1.

EXERCISES

8.1. Prove that both alternatives (8.1), (8.2) can hold in the same group G for different choices of r.

8.2. Let $G = \langle a, b \mid ab = ba \rangle \cong Z^2$ and $a' = a^i b^j$, $b' = a^k b^l \in G$, $i, j, k, l \in \mathbb{Z}$. Prove that a', b' generate G if and only if $\det \begin{pmatrix} i & j \\ k & l \end{pmatrix} = \pm 1$.

8.3. Prove that G_1^3 and G_2^4 contain no reflections.

8.4. Prove that $G_1^1 \not\cong G_1^2$ and deduce from this and the previous exercise that no two of the four G_1-groups can be isomorphic.

8.5. By computing their derived factor groups, show that the five G_2-groups contain at least three isomorphism types. Can you prove a stronger result?

8.6. List the elements of each possible finite order in each of the three G_4-groups.

8.7. Find two elements of G_4^2 that generate the whole group and write down a presentation for it on these two generators.

8.8. Show that in G_3^2 all centres of rotation lie on axes of reflection, but that this is not the case in G_3^1.

8.9. The subgroup $H = \langle a, b, s^2 \rangle$ of G_6^1 is another crystallographic group. Which one?

8.10. In (at least) eight of the 12 crystallographic groups G with $|G : G^+| = 2$ the translation subgroup has a complement C. Identify C in each case.

8.11. It turns out that no two of the 17 plane crystallographic groups are isomorphic. This can be proved by comparing such group invariants as the index of the derived group and the possible finite orders for the elements. Filling in the details would make a nice project.

8.12. Another nice project would be to decide which of the 17 groups is isomorphic to a subgroup of which of the others.

Tessellations of the Plane

A **tessellation of the plane** is a covering of $\mathbb{R}^2$ by non-overlapping polygons. By "covering" here we mean that every point of $\mathbb{R}^2$ lies in (the interior of) or on (the boundary of) at least one of the polygons. By "non-overlapping" we mean that two distinct polygons intersect, if at all, in a part of the boundary of each. Our polygons thus have a boundary and an interior, and we think of them as tiles, or tessera. For the sake of convenience, we make the additional assumption that all interior angles of all polygons are less than π. This is equivalent to the stipulation that each polygon is the intersection of (at least three) half-planes, and is therefore convex. Another consequence is that the intersection of any two polygons is a connected set.

9.1 Regular Tessellations

A polygon is said to be **regular** if its interior angles are all equal and its sides are all of equal length. A tessellation by congruent regular polygons, any two of which meet, if at all, in a common side or vertex of each, is said to be **regular**. In a regular tessellation by n-gons meeting m at a vertex, $m, n \geq 3$, the angle-sum of each polygon is $(n - 2)\pi$. Each interior angle is thus $(n - 2)\pi/n$, and since there are m of these at each vertex, we have

$$m(n - 2)\pi/n = 2\pi$$
$$\Rightarrow \ m(n - 2) = 2n$$

$$\Rightarrow \ mn - 2m - 2n = 0$$
$$\Rightarrow \ (m-2)(n-2) = 4.$$

Since there are only three ways of writing 4 as a product of two positive integers, there are at most three types of regular tessellation of $\mathbb{R}^2$, those with parameters

$$(m, n) = (4, 4), \ (3, 6), \ (6, 3).$$

Happily, these all exist: see Figs. 19.1–19.3.

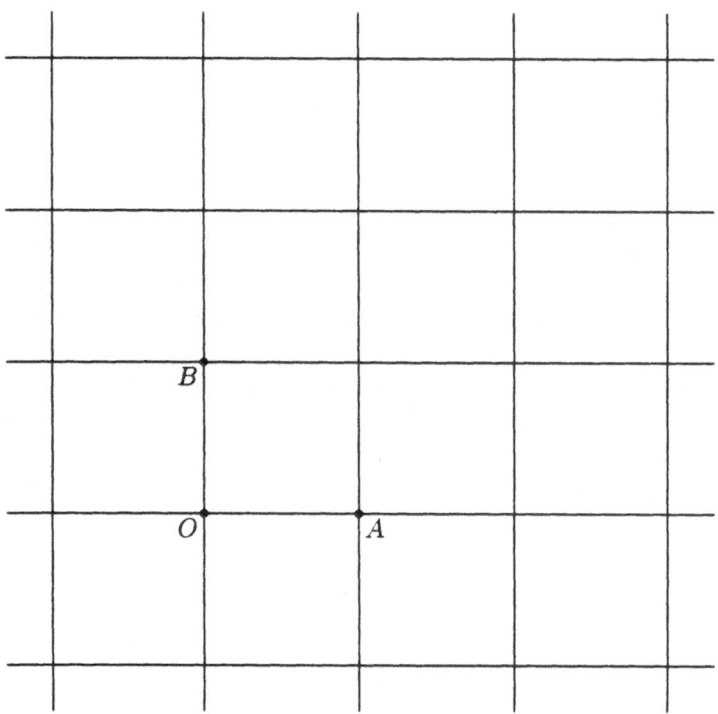

Figure 9.1 $(4, 4)$: squares meeting in fours

We compute the symmetry groups of these figures using the reference points O, A, B as marked. In each case we take $a = t(\overrightarrow{OA})$, $b = t(\overrightarrow{OB})$. In $(4, 4)$ we take $r = r(l)$, where $l = l(O, A)$. Then

$$ab = ba, \ r^2 = 1, \ a^r = a, \ b^r = b^{-1}. \tag{9.1}$$

Taking r to be reflection in the angle bisector of OA, OB in $(3, 6)$ and $(6, 3)$, we get

$$ab = ba, \ r^2 = 1, \ a^r = b, \ b^r = a. \tag{9.2}$$

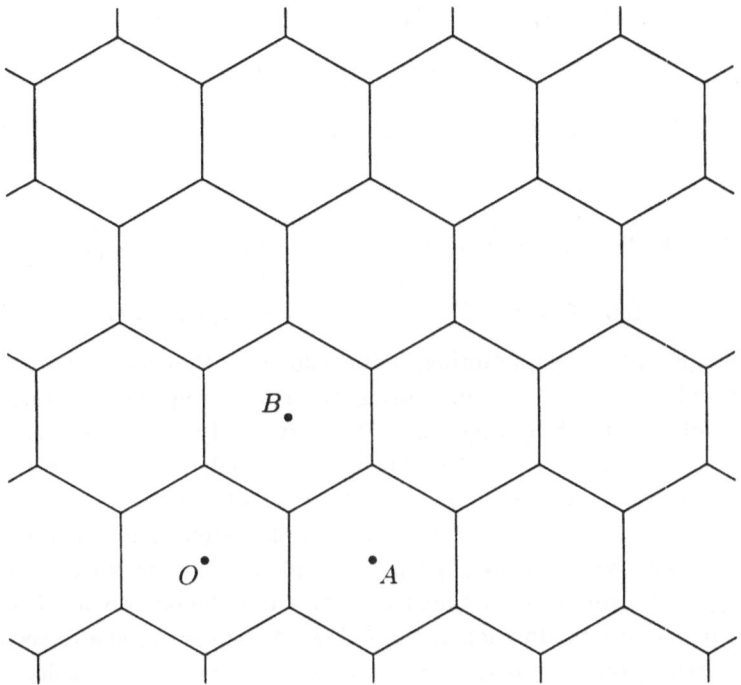

Figure 9.2 $(3,6)$: hexagons meeting in threes

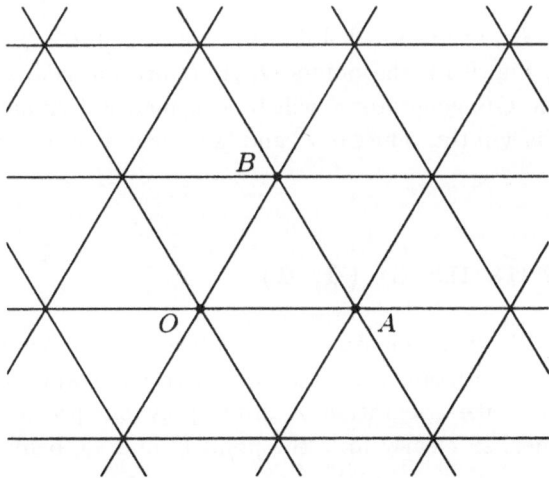

Figure 9.3 $(6,3)$: triangles meeting in sixes

Letting $s = s(O, \pi/2)$ in $(4, 4)$, we get

$$s^4 = 1, \ a^s = b, \ b^s = a^{-1}, \ (sr)^2 = 1. \tag{9.3}$$

With $s = s(O, \pi/3)$ in $(3, 6)$ and $(6, 3)$, we have

$$s^6 = 1, \ a^s = b, \ b^s = a^{-1}b, \ (sr)^2 = 1. \tag{9.4}$$

Since relations (9.1), (9.3) define G_4^1 and (9.2), (9.4) define G_6^1, we have shown that

$$\text{Sym}(4, 4) = G_4^1, \quad \text{Sym}(3, 6) = \text{Sym}(6, 3) = G_6^1.$$

Although rather disappointing, it is no surprise that the symmetry groups of $(3, 6)$ and $(6, 3)$ are the same, for the corresponding tessellations are dual to one another in the following sense. When two polygons abut along an edge, join their centres by a new edge. Then the new edges define the tessellation dual to the original one. It is clear that $(4, 4)$ is self-dual.

Having dealt with G_4^1 and G_6^1, we now seek tessellations corresponding to the other plane crystallographic groups. To this end, we relax the conditions of regularity to the single requirement that all the polygons involved in the tessellation be congruent. In each of the following figures, we shall specify translations t, reflections r and glide reflections q in terms of an ordered pair of distinct points $C, D \in \mathbb{R}^2$ as follows:

$t(C, D)$ is the translation that takes C to D,

$r(C, D)$ is reflection in the line through C and D,

$q(C, D) = r(C, D)t(C, D)$.

Also, for groups G_n of type $n = 2, 3, 4, 6$, we take $s = s(O, 2\pi/n)$. In every figure except the last, Fig. 9.18, the points O, A, B are marked, and $a = t(O, A)$, $b = t(O, B)$. The OR generator r will be specified individually in every case where it exists, as will the values of r^2 and $(sr)^2$ where necessary or convenient.

9.2 Descendants of (4, 4)

In Figs. 9.4 and 9.5, every square of $(4, 4)$ is bisected chessboardwise as shown into a pair of 2×1 rectangles, a pair of congruent trapezia respectively. In the latter case, the like orientation of all the polygons precludes the existence of r. In the former, sr clearly fixes the point C and so, being OR, must be a reflection.

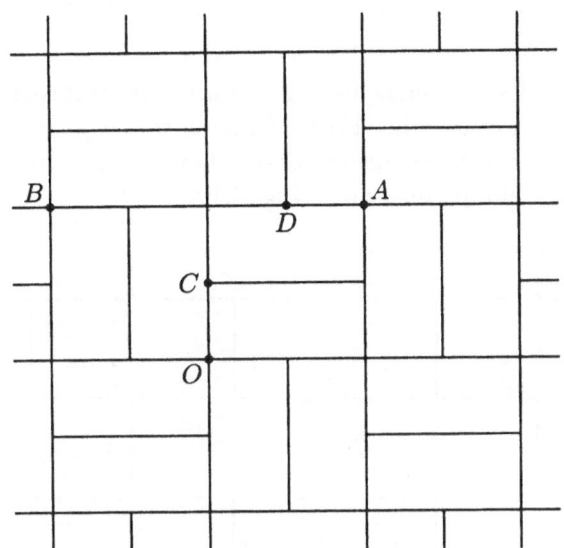

Figure 9.4 G_4^2: $r = q(C, D)$, $r^2 = a$, $(sr)^2 = 1$

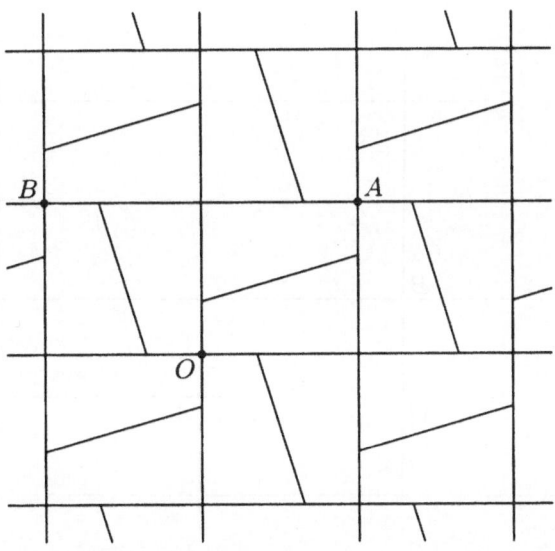

Figure 9.5 G_4

9.3 Bricks

The 2×1 rectangles that make up Fig. 9.4 are a fruitful source of examples. They can be laid in courses like bricks, either correctly as in Fig. 9.6, or as in Figs. 9.7–9.9 with crossed, staggered, zig-zag bonds respectively. They can also be treated as tiles in a parquet floor (Fig. 9.10).

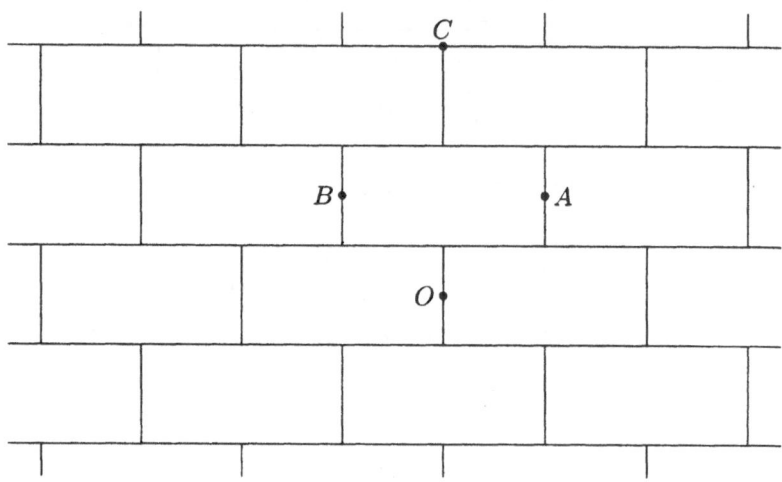

Figure 9.6 G_2^3: $r = r(O, C)$, $r^2 = (sr)^2 = 1$

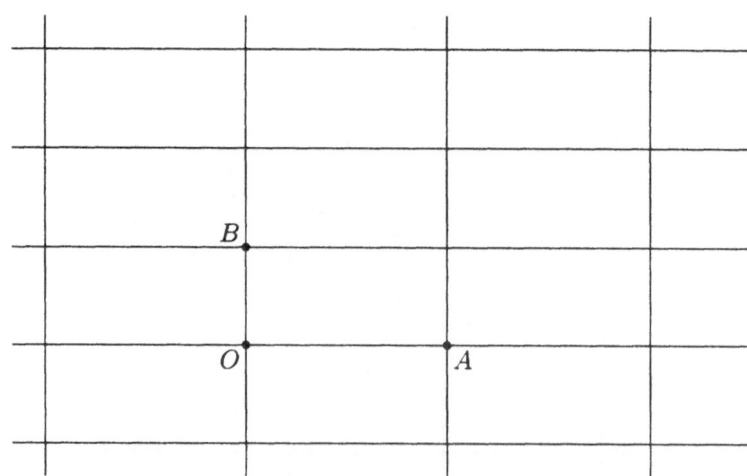

Figure 9.7 G_2^1: $r = r(O, A)$, $r^2 = (sr)^2 = 1$

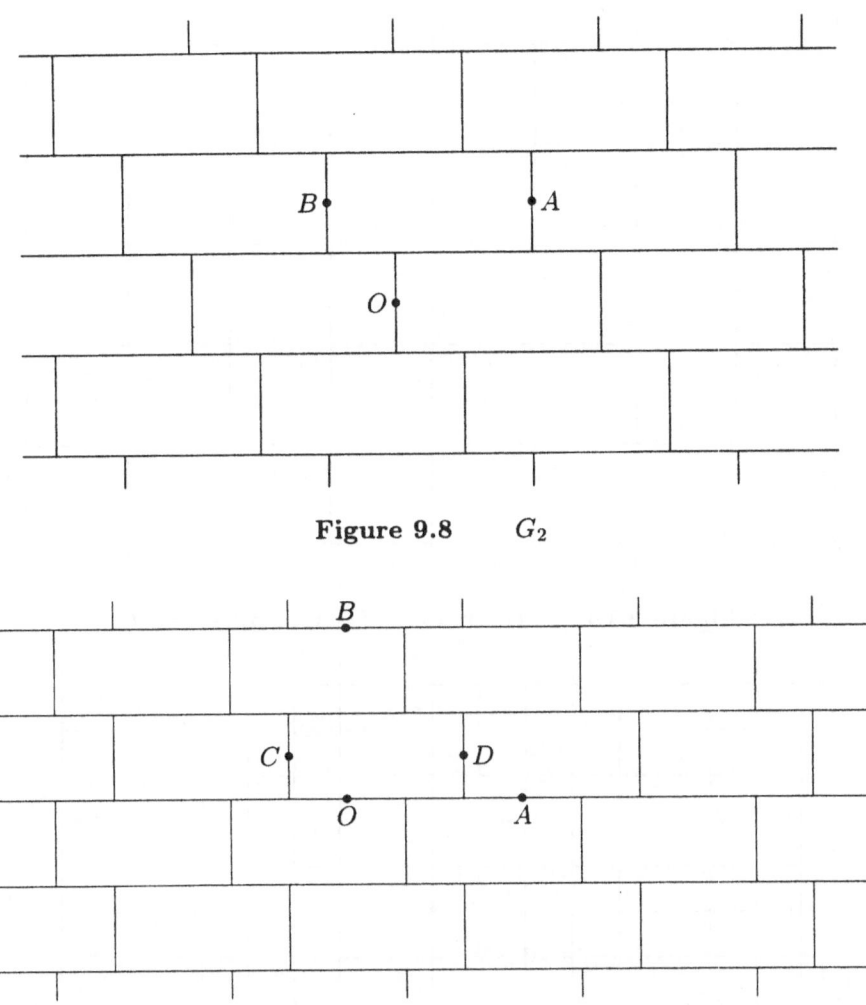

Figure 9.8 G_2

Figure 9.9 G_2^2: $r = r(C, D)$, $r^2 = 1$, $(sr)^2 = b$

9.4 Split Bricks

A horizontal 6×1 rectangle can be divided by vertical segments into six unit squares, each of which can be bisected by a vertical or horizontal segment. The bi-infinite sequence $\ldots hvhhvv \ldots$ of period 6 is not a translate of its reversal. So the rotational symmetry in Figs. 9.6 and 9.7 is lost if we replace each brick by a 6×1 strip of this form, as in Figs. 9.11 and 9.12. Symmetries are also lost when the bricks in Fig. 9.9 are bisected by diagonals as shown in Figs. 9.13 and 9.14. In the former, oppositely orientated courses alternate (lr) and in the latter, they have period 3 (rrl).

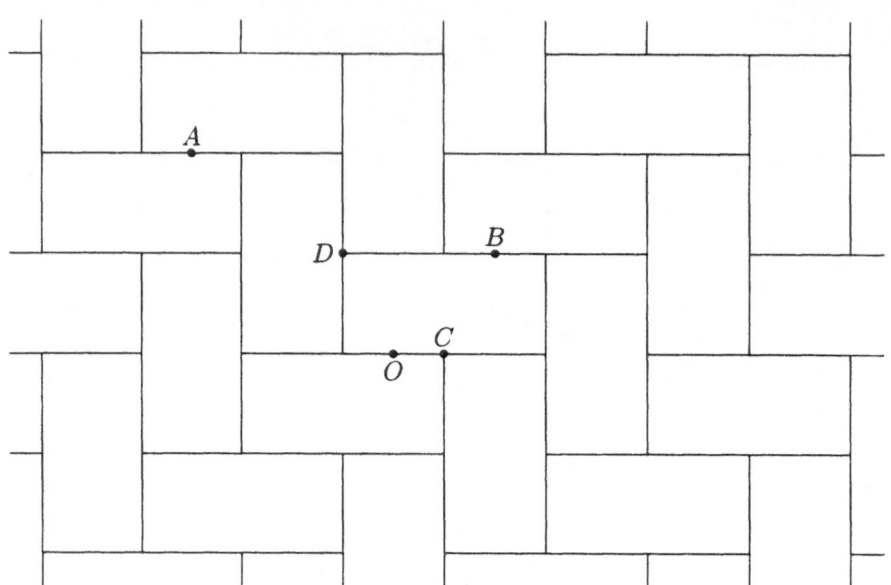

Figure 9.10 G_2^4: $r = q(C, D)$, $r^2 = a$, $(sr)^2 = b$

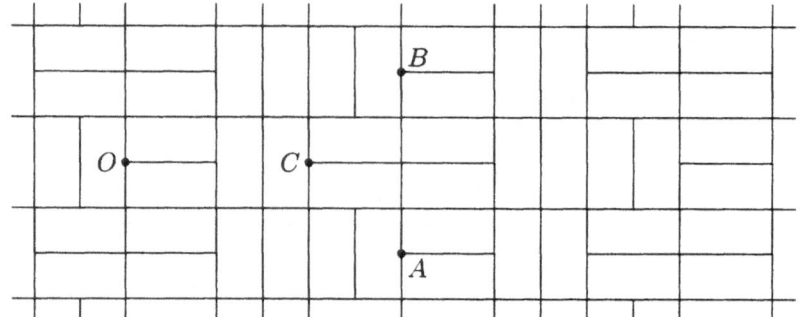

Figure 9.11 G_1^2: $r = r(O, C)$, $r^2 = 1$

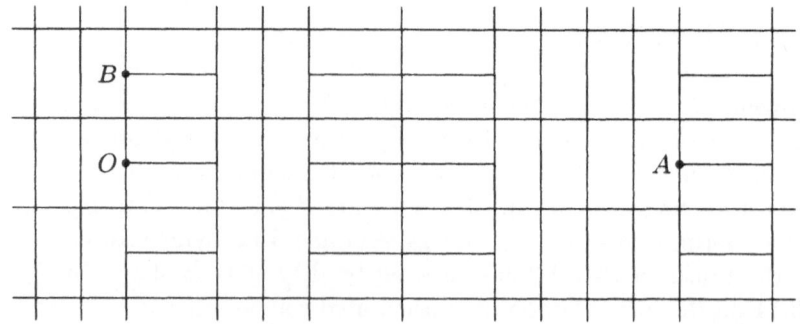

Figure 9.12 G_1^1: $r = r(O, A)$, $r^2 = 1$

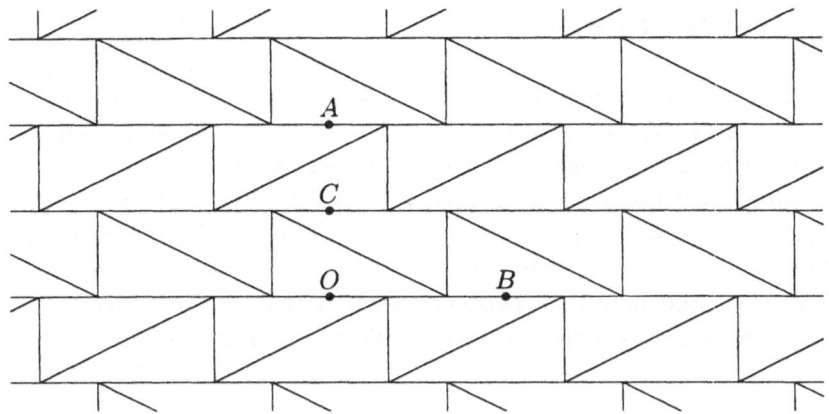

Figure 9.13 G_1^3: $r = q(O, C)$, $r^2 = a$

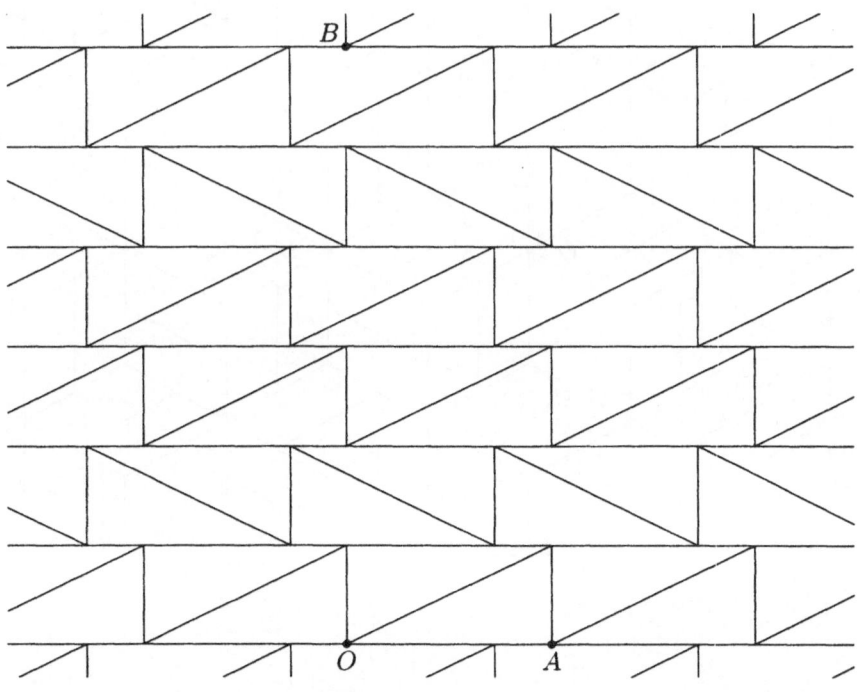

Figure 9.14 G_1

9.5 Descendants of (3, 6)

Each of the hexagons in $(3, 6)$ can be trisected by a rocket, either taking off λ or landing $\curlyvee$. We take two such subdivisions, homogeneous (all λ) and periodic (two-thirds λ, one-third $\curlyvee$), and bisect all the resulting rhombs into like-oriented parallelograms as in Figs. 9.15 and 9.16, respectively.

Each hexagon in $(3, 6)$ can also be bisected into a pair of trapezia in three different ways. Sections of two such subdivisions are depicted in Figs. 9.17 and 9.18, the essential difference between which is that in the former there are centres of rotation not lying on axes of reflections, and in the latter there are not.

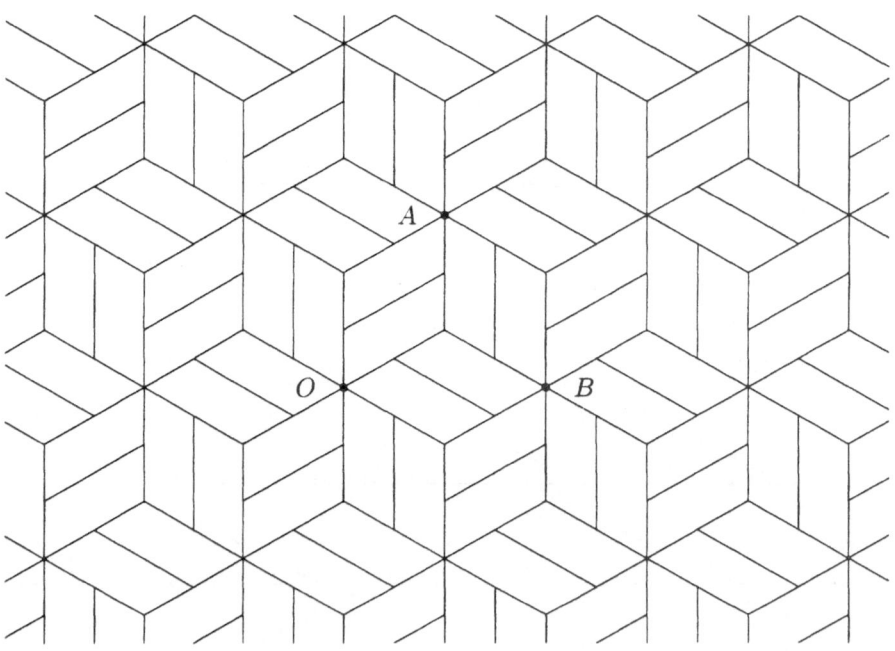

Figure 9.15 G_6

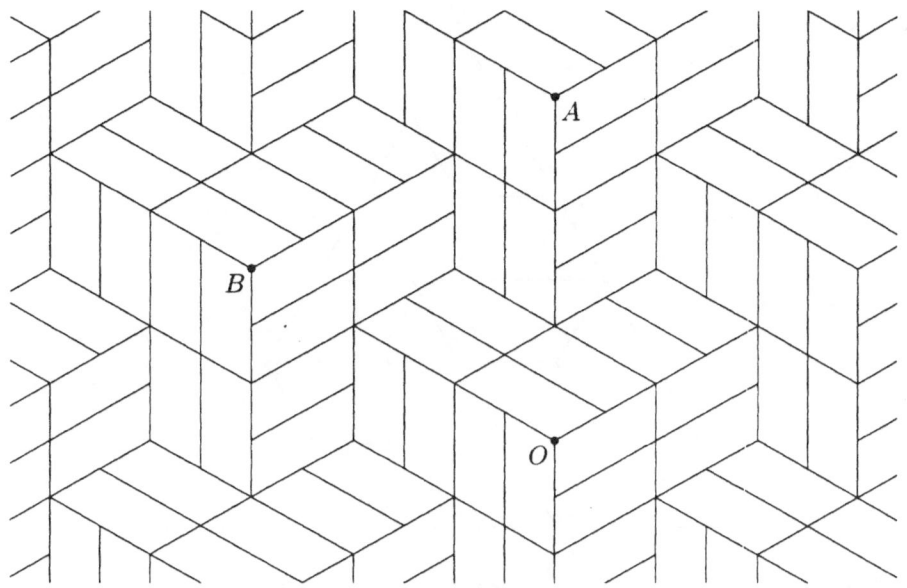

Figure 9.16 G_3

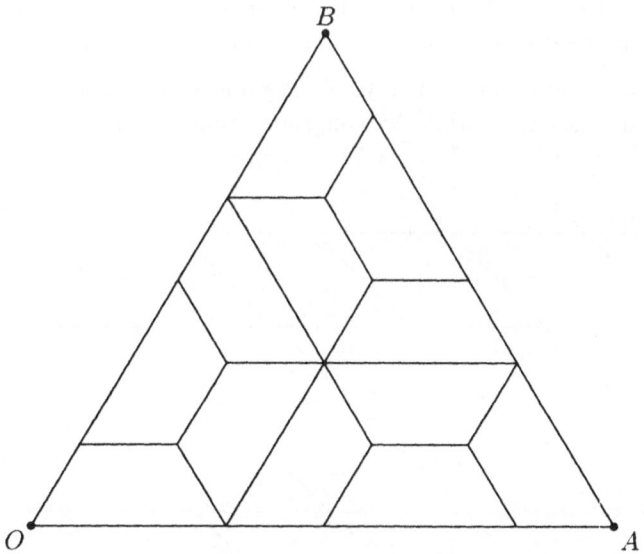

Figure 9.17 G_3^1: $r = r(O, A)$, $sr = r(O, Bs^{-1})$

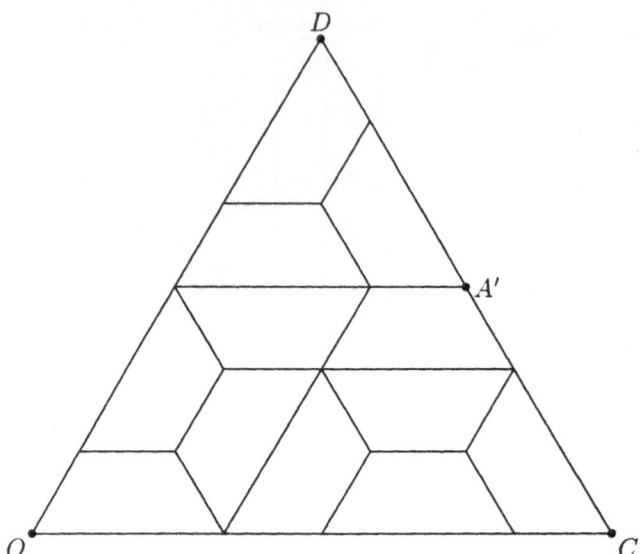

Figure 9.18 G_3^2: $r = r(O,C)$, $a = t(O,A')^2$, $b = a^r$, $sr = r(O,D)$

EXERCISES

9.1. Find figures simpler than those above for G_3, G_1, G_1^1, G_1^2.

9.2. Identify the symmetry groups of the eight figures below among the 17 plane crystallographic groups (only two of them are the same).

9.3. Prove that every plane crystallographic group is the symmetry group of a tessellation of $\mathbb{R}^2$ by congruent quadrilaterals.

Figure 9.19   Exercise 9.2a

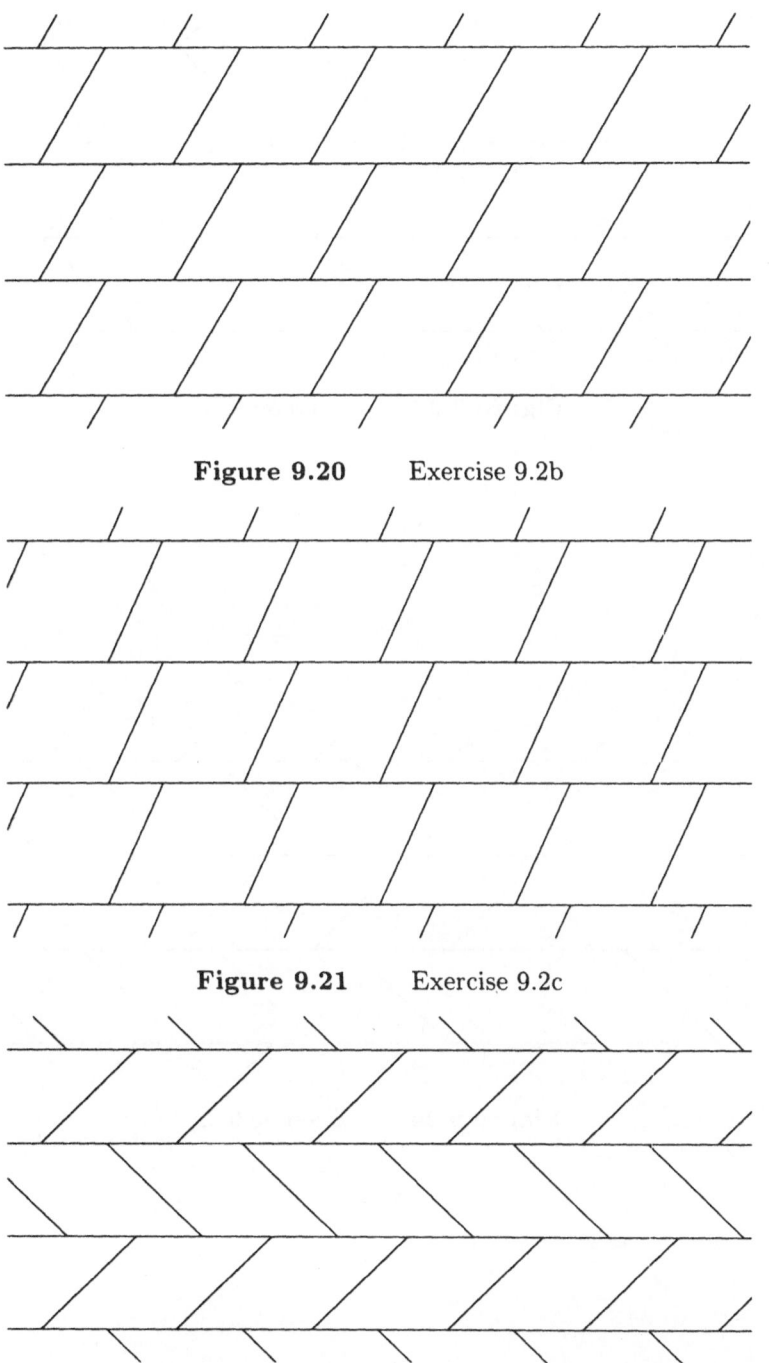

Figure 9.20   Exercise 9.2b

Figure 9.21   Exercise 9.2c

Figure 9.22   Exercise 9.2d

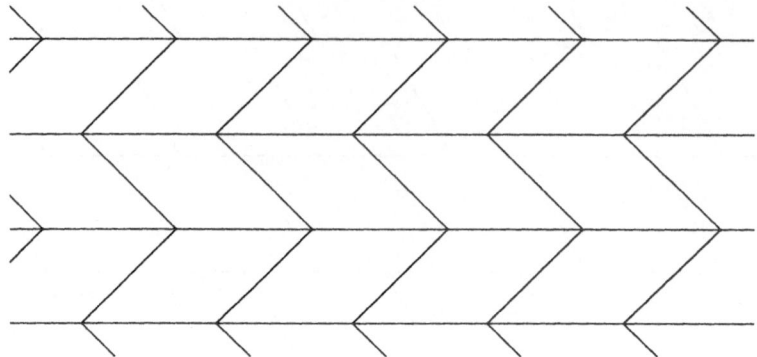

Figure 9.23   Exercise 9.2e

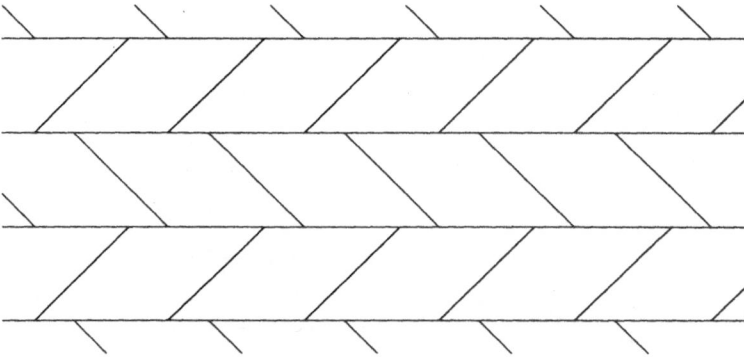

Figure 9.24   Exercise 9.2f

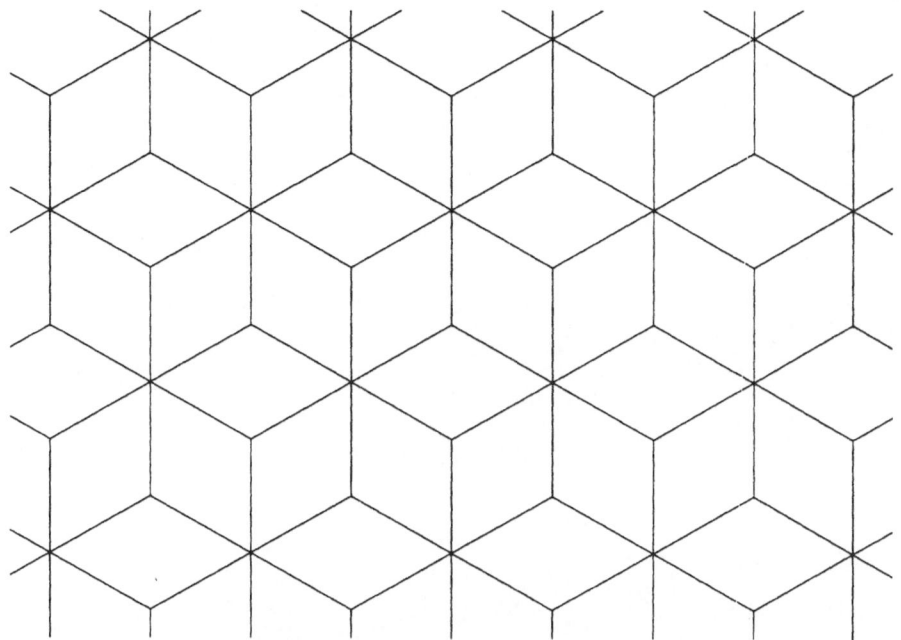

Figure 9.25   Exercise 9.2g

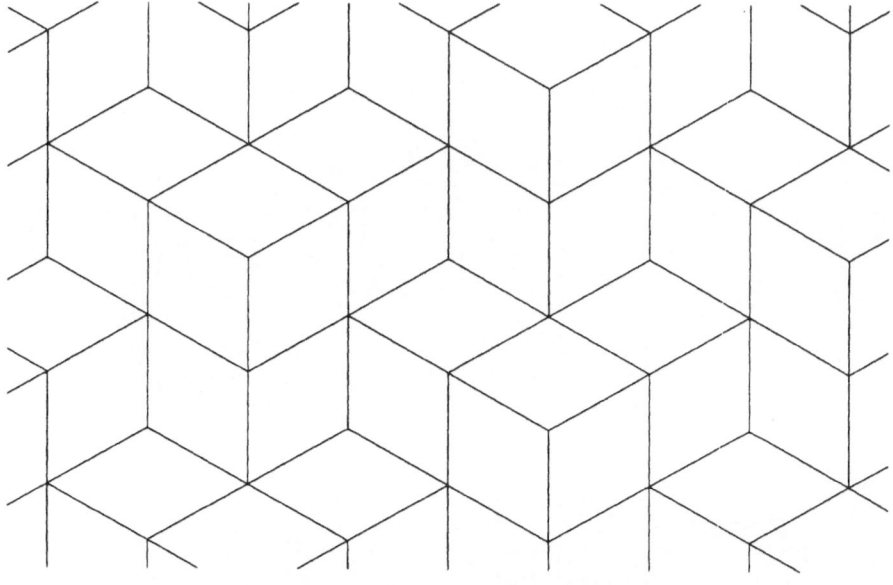

Figure 9.26   Exercise 9.2h

10

Tessellations of the Sphere

A **sphere** is the set of all points at a fixed distance from a given point. The given point is called the **centre**, and the fixed distance the **radius**, of the sphere. The points are those of any set on which distance is defined, that is, of a metric space. In this chapter we take this to be Euclidean 3-space with the Pythagorean metric. Sensibly letting the centre be the origin of coordinates and the radius be the unit of length, we get the **unit sphere**

$$\mathbb{S}^2 = \{(x, y, z) \in \mathbb{R}^3 \mid x^2 + y^2 + z^2 = 1\}.$$

The surface of the earth resembles a sphere and is traditionally parametrised by latitude and longitude. It is sometimes helpful to think in such geographical terms, and others like "antipodes" and "equator." We shall not hestitate to make use of such nomenclature in what follows.

10.1 Spherical Geometry

While geometry on $\mathbb{S}^2$ has many features in common with plane geometry, there are a number of important differences. Thus, the basic objects of study are the same: points, lines, angles, polygons, and so on. Also, some of their properties, such as

(a) there is a line through any two points,

(b) vertically opposite angles are equal

are valid in both geometries. On the other hand, other properties like

(c) given a line l and a point P not on l, there is a unique line through P parallel to l,

(d) the angle-sum of a triangle is equal to π

fail in the spherical case. In what follows, we shall capitalise on or make allowances for such differences according as they are convenient or not.

We proceed to give a fairly informal description of the points, lines, angles and polygons in spherical geometry. The first of these is easy: **points** are just elements of $\mathbb{S}^2$.

The **lines** are the intersection with $\mathbb{S}^2$ of planes through the centre, and are often called great circles. Thus, lines of longitude are great semicircles, and the equator is a great circle but other lines of latitude are not. Since three non-collinear points in $\mathbb{R}^3$ determine a plane, there is a unique line in $\mathbb{S}^2$ through a pair of non-antipodal points P and Q. P and Q are thus joined by two line segments, or arcs, of unequal length. The shorter of these is the path between P and Q in $\mathbb{S}^2$ of minimal length: the **geodesic**. This defines the distance between P and Q as the (smaller) angle $P\widehat{O}Q$, and it can be proved that this metric coincides with that inherited from the Pythagorean metric on $\mathbb{R}^3$. If on the other hand P and Q are antipodal points, every arc between them has the same length, namely π, and this is therefore the maximal distance possible in $\mathbb{S}^2$.

Two great circles C_1, C_2 in $\mathbb{S}^2$ meeting at a point P pass under central projection to two ordinary lines l_1, l_2 in the plane tangent to $\mathbb{S}^2$ at P. The **angle** between C_1 and C_2 is then defined as the usual planar angle between l_1 and l_2 at P. This is the same as the angle between the planes containing C_1 and C_2. Some care is needed here as there are actually four angles at P: two vertically opposite pairs. We make the convention that all angles are positive, and distinguish one from an adjacent supplementary pair by the context.

Turning finally to polygons, we have already created some: two great circles divide $\mathbb{S}^2$ into four digons, or **lunes**. These degenerate but useful objects have no proper analogue in plane geometry. Note that every lune is regular: it is bounded by two semicircles, both of length π, meeting in two antipodal points at equal angles. The lune is determined up to congruence by this angle.

As in the planar case, we wish our general n-gon to have n vertices joined cyclically by n arcs which together form a simple closed curve. But in the spherical case we require more than this, to avoid pathology of the following kind. Take the case $n = 3$. In the plane, three non-collinear points determine a triangle. Not so on the sphere: a side could be replaced by the complementary arc of the same great circle. Nor is it sufficient to specify the three sides, which bound two complementary triangles in $\mathbb{S}^2$. We get around these and other difficulties by making the following definition.

A **polygon** on $\mathbb{S}^2$ is an intersection of hemispheres. When n hemispheres are involved without redundancy the polygon will have n vertices and n sides, and will be called an n-gon. In deference to common usage, the terms lune, triangle, quadrilateral are employed in the cases $n = 2, 3, 4$, respectively, the first of these being regarded as degenerate. There follows some discussion of this definition and its consequences.

The definition is certainly slick and, while it may appear rather restrictive, it is sufficient for our purposes. It is also a very natural one geometrically in the sense of the following general notion, defined here for $\mathbb{S}^2$. A subset C of $\mathbb{S}^2$ is said to be **convex** if, for any two points P and Q of C, the geodesic joining P and Q lies entirely within C. (Note that care is needed in the case when P and Q are antipodal, but for n-gons this can only happen when $n = 2$.) Since (open) hemispheres are convex, and intersections of convex sets are convex, it follows that our polygons are convex. This is the first assertion of the following theorem.

Theorem 10.1

Let T be an n-gon on $\mathbb{S}^2$. Then

(a) *T is convex,*

(b) *each interior angle of T is less than π,*

(c) *any pair of non-adjacent vertices of T are joined by a geodesic arc that partitions T into two polygons.*

Proof

(a) is proved above and (b) is obvious (two great circles intersect at an angle less than π). As for (c), which applies only when $n \geq 4$, let P and Q be the vertices and a the geodesic arc joining them. Then $a \subseteq T$ as T is convex and, if H_1 and H_2 are the hemispheres bounded by the great circle containing a, it is clear that $T = (T \cap H_1) \cup (T \cap H_2)$ is the required partition. $\square$

10.2 The Spherical Excess

As mentioned in the previous section, the angle-sum of a spherical triangle is not equal to π. In fact, it always exceeds π. This can be read off from the much more general and precise result (Theorem 10.2) proved below. The key idea is the subject of the following general definition for $n \geq 2$: the **spherical excess** ε of an n-gon on $\mathbb{S}^2$ is the amount by which its angle-sum σ exceeds that of a plane n-gon,

$$\varepsilon = \sigma - (n-2)\pi.$$

We claim that ε is always positive, which agrees with the above when $n = 3$ and is clear when $n = 2$: the spherical excess of a lune is just twice the lunar angle. A crucial fact is the close relationship that exists between the spherical excess of a polygon and its area. It could hardly be closer.

Theorem 10.2

Let T be an n-gon on $\mathbb{S}^2$ with area A and spherical excess ε. Then $A = \varepsilon$.

Proof

This is divided into three cases: $n = 2$, $n = 3$, $n \geq 3$. The critical case $n = 3$ is known as Girard's theorem.

(a) $n = 2$. It is intuitively clear, and easily shown using calculus (Exercise 10.4), that the area of a lune T is proportional to its angle α. Since the hemisphere has lunar angle π and area 2π, the constant of proportionality must be 2. So the area of T is 2α, which equals the spherical excess.

(b) $n = 3$. Take a triangle T with vertices P, Q, R and angles α, β, γ as shown in Fig. 10.1. Let P', Q', R' be antipodal to P, Q, R respectively, and assume that T has area A. Now the hemisphere bounded by $PQP'Q'$ and containing R is partitioned into four triangles

$$PQR,\ QP'R,\ P'Q'R,\ Q'PR$$

of areas, respectively,

$$A,\ 2\alpha - A,\ 2\gamma - A,\ 2\beta - A,$$

using part (a) and the fact that the triangles PQR, $P'Q'R'$ are congruent (Exercise 10.5). Since the area of the hemisphere is 2π, we get

$$2\pi = 2(\alpha + \beta + \gamma) - 2A,$$

that is,

$$A = \alpha + \beta + \gamma - \pi = \varepsilon.$$

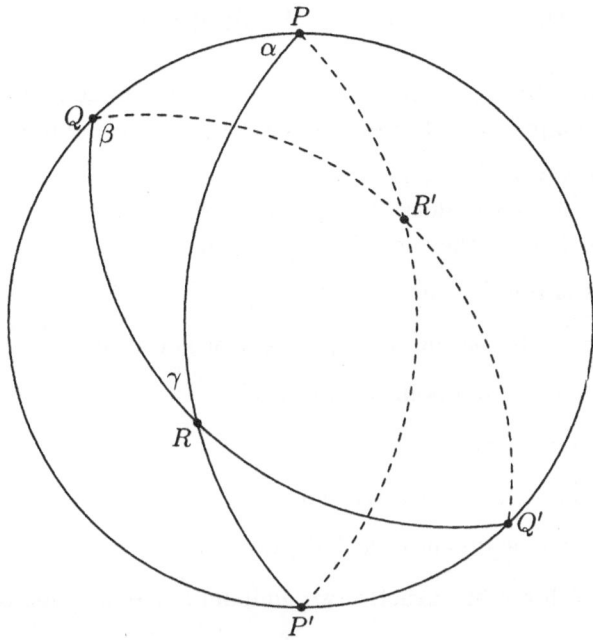

Figure 10.1

(c) $n \geq 4$. This is by induction on $n \geq 3$ based on part (b). So take an n-gon T with $n \geq 4$ and assume the result for m-gons with $3 \leq m < n$. In accordance with Theorem 10.1(c), divide T into an l-gon T_1, and an m-gon T_2 by drawing a geodesic between two non-adjacent vertices. If T, T_1, T_2 have areas A, A_1, A_2, angle-sums σ, σ_1, σ_2 and spherical excesses ε, ε_1, ε_2, respectively, it is clear that

$$A = A_1 + A_2, \quad \sigma = \sigma_1 + \sigma_2.$$

Further, since $l + m = n + 2$,

$$(n-2)\pi = (l-2)\pi + (m-2)\pi.$$

Finally, since $l, m < n$,

$$A_1 = \varepsilon_1, \quad A_2 = \varepsilon_2$$

by the inductive hypothesis, whence

$$A = A_1 + A_2 = \varepsilon_1 + \varepsilon_2$$
$$= \sigma_1 - (l-2)\pi + \sigma_2 - (m-2)\pi$$
$$= \sigma - (n-2)\pi = \varepsilon. \qquad \square$$

10.3 Tessellations of the Sphere

A **tessellation** of $\mathbb{S}^2$ is a covering by a finite number of non-overlapping polygons. To be more precise, a tessellation consists of three finite sets,

$\mathcal{T}$ of polygons whose union is $\mathbb{S}^2$,
$\mathcal{A}$ of arcs that are the sides of polygons in $\mathcal{T}$,
$\mathcal{P}$ of points that are the endpoints of arcs in $\mathcal{A}$,

satisfying rules of **incidence**,

(a) two polygons intersect in an arc, a point or not at all;

(b) two arcs intersect in a point or not at all;

and of **non-degeneracy**,

(c) each polygon has at least three sides;

(d) each point is an endpoint of at least three arcs.

Note that each arc has exactly two endpoints and is a side of exactly two polygons.

In this context, the points, arcs and polygons are usually referred to as the **vertices**, **edges** and **faces**, respectively, of the tessellation. We therefore label them

$$\mathcal{P} = \{P_i \mid 1 \leq i \leq v\}, \quad \mathcal{A} = \{A_1 \mid 1 \leq i \leq e\}, \quad \mathcal{T} = \{T_i \mid 1 \leq i \leq f\}.$$

Now let P_i have n_i sides, angle-sum σ_i and spherical excess ε_i, $1 \leq i \leq f$:

$$\varepsilon_i = \sigma_i - (n_i - 2)\pi. \tag{10.1}$$

Since counting all the sides of all the faces counts the edges twice each, we have

$$\sum_{i=1}^{f} n_i = 2e. \tag{10.2}$$

Since the angle-sum at each vertex is 2π, adding all the angles in all the polygons gives

$$\sum_{i=1}^{f} \sigma_i = 2\pi v. \tag{10.3}$$

We now sum (10.1) from 1 to f and use Theorem 10.2, (10.3) and (10.2) to get

$$4\pi = \sum_{i=1}^{f} \varepsilon_i = \sum_{i=1}^{f} \sigma_i - \pi \sum_{i=1}^{f} n_i + 2\pi \sum_{i=1}^{f} 1$$
$$= 2\pi v - 2\pi e + 2\pi f.$$

This delivers the following famous result.

Theorem 10.3 (Euler's formula)

If T is a tessellation of $\mathbb{S}^2$ with v vertices, e edges and f faces, then

$$v - e + f = 2. \tag{10.4}$$

$\square$

We now specialise to the case of regular tessellations, defined as follows. An n-gon on $\mathbb{S}^2$ is **regular** if all its n angles are of equal magnitude and all its n sides are of equal length. A tessellation T of $\mathbb{S}^2$ is **regular** if all its polygons are regular and congruent. The last condition implies that every polygon has the same number, say n, of sides, and the same number, say m, of polygons meet at every vertex. It follows from Theorem 10.2 that a regular tessellation T is determined by the pair (m, n). We assume here that $m, n \geq 3$ but bear in mind the degenerate case when m or $n = 2$.

Now with v, e, f as above we can simplify (10.2) and (10.3) as follows:

$$fn = 2e, \quad fn = mv, \tag{10.5}$$

since $\sigma_i = n2\pi/m$, $1 \leq i \leq f$. Thus, $mv = fn = 2e$, and we can eliminate f and v from (10.4):

$$\frac{2e}{m} - e + \frac{2e}{n} = 2, \tag{10.6}$$

that is,

$$\frac{1}{m} + \frac{1}{n} = \frac{1}{2} + \frac{1}{e} > \frac{1}{2}.$$

It follows that m and n cannot both exceed 3, and when one of them is equal to 3, the other is at most 5. This shows that there are at most five possibilities for T.

Theorem 10.4

If T is a regular tessellation of $\mathbb{S}^2$ with parameters $m, n \geq 3$, then (m, n) is one of

$$(3, 3), \quad (3, 4), \quad (3, 5), \quad (4, 3), \quad (5, 3).$$
$\square$

The happy fact that all five possibilities for a regular tessellation of $\mathbb{S}^2$ can actually be realised in $\mathbb{R}^3$ has essentially been known from antiquity. We shall construct and describe them in the next section. To get some idea of what we are looking for, we calculate from (10.5) and (10.6) the values of v, e, f in each case, entering their values in Table 10.1. The meaning of the symbols in the first and last columns will be explained in the next section.

Table 10.1

	m	n	v	e	f	g
T	3	3	4	6	4	12
C	3	4	8	12	6	24
O	4	3	6	12	8	24
D	3	5	20	30	12	60
I	5	3	12	30	20	60

10.4 The Platonic Solids

We begin with a few informal words about the close, and fairly obvious, correspondence between regular tessellations of $\mathbb{S}^2$ and regular solids in $\mathbb{R}^3$, beginning with a definition of the latter.

A **solid**, or convex polyhedron, in $\mathbb{R}^3$ is a bounded intersection of a finite number of half-spaces. Its boundary consists of (plane) polygonal faces bounded by edges and vertices satisfying the rules (a)–(d) of incidence and non-degeneracy in the above definition (of tesselation of $\mathbb{S}^2$) with "arc" replaced by "edge." A plane polygon is **regular** if, as above, its sides all have the same length and its angles the same magnitude, and a solid is **regular** if its faces are all regular and congruent and the same number of faces meet at every vertex.

Now let $\mathcal{T}$ be a regular tessellation of $\mathbb{S}^2$ with f faces. Since each face is regular, all its vertices lie in a plane in $\mathbb{R}^3$ that bounds the half-space containing the centre O of $\mathbb{S}^2$. The intersection S of these f half-spaces is contained in $\mathbb{S}^2$ and so is bounded, and rules (a)–(d) are inherited, as are the regularity and congruence of the faces. That the correspondence of $\mathcal{T}$ and S is one-to-one follows by central projection. Another way of getting S from $\mathcal{T}$ is by taking the convex hull in $\mathbb{R}^3$ of its vertex set $\mathcal{P}$.

To construct the regular solids, we briefly stoop to choosing coordinates. Thus, let x, y, z denote the usual Cartesian coordinates of $\mathbb{R}^3$ with origin O. Then each of the five regular solids can be specified by giving its vertices, as follows.

The easiest to specify and most familiar is the **cube C**, with eight vertices $(\pm 1, \pm 1, \pm 1)$. (To get inside $\mathbb{S}^2$, divide each coordinate of every vertex by $\sqrt{3}$.)

Of these, the four points with an even number (0 or 2) of positive coordinates define the **tetrahedron T**. Likewise the other four, and these two tetrahedra intersect in the **octahedron O**, whose vertices $(\pm1,0,0)$, $(0,\pm1,0)$, $(0,0,\pm1)$ are the centres of the six faces of **C**: we then say that **O** is the **dual** of **C**.

The tetrahedron is a triangular pyramid (the base is a triangle), regular in the sense that all six edges have the same length. The octahedron consists of two regular square pyramids with the bases glued together. Now take two regular pentagonal pyramids and put their bases together. Rotate one of them through $\pi/5$ about the line l joining their apices and then separate them by translation along l until each base vertex of one together with the two nearest base vertices of the other form the three vertices of an equilateral triangle. Inserting these 10 equilateral triangles into the figure, we get the **icosahedron I**.

The regular **dodecahedron D** can be constructed from the cube as follows. Take a regular plane pentagon $ABCDE$ of unit side and draw in a diagonal, length τ say, which dives the pentagon into a trapezium $BCDE$ and an isosceles triangle ABE. Now place two such trapezia with their bases BE on opposite side of a square $BEE'B'$ of side τ at such an inclination that their sides CD meet as shown in Fig. 10.2. The triangular ends DEE', $CB'B$ can be nicely filled in with two copies of $\triangle ABE$. Now perform this construction on each of the six faces of a cube of side τ, taking care that the "ridges" CD on any pair of adjacent faces are skew (not coplanar). Et voilà!

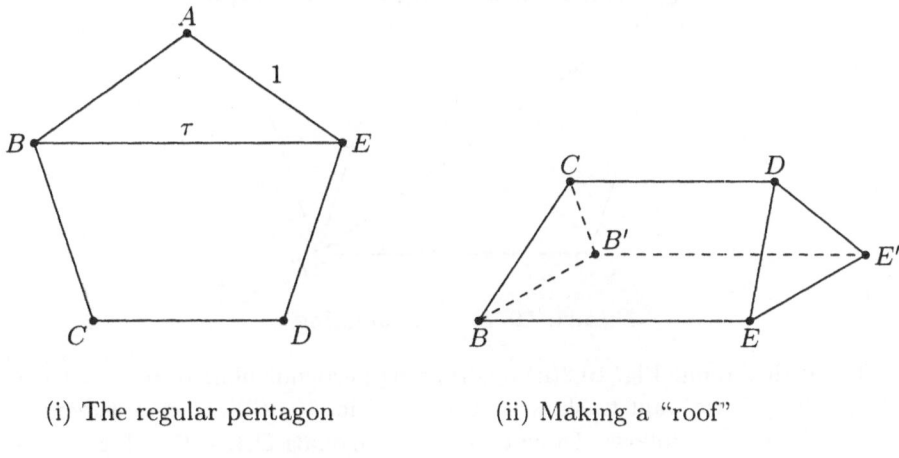

(i) The regular pentagon   (ii) Making a "roof"

Figure 10.2   Construction of **D**

But there's a catch. While we do indeed have a figure bounded by 12 pentagons (one for each edge of the cube) meeting in threes at each vertex, it remains to be checked that the trapezium and triangle meeting at each edge of the cube are in fact *coplanar*.

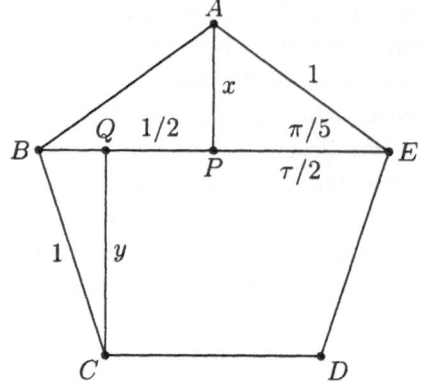

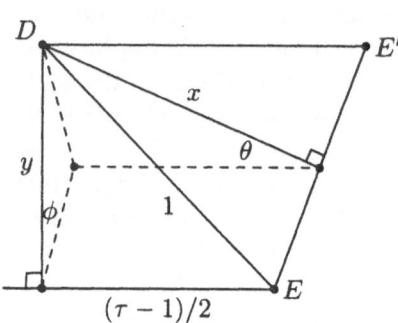

(i) Defining x and y (ii) $\theta + \phi = \pi/2$

Figure 10.3   Checking coplanarity

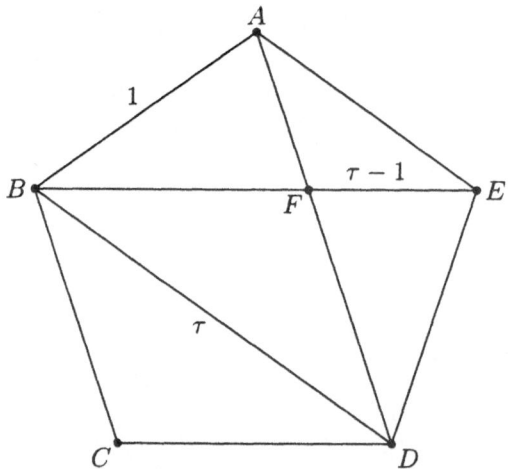

Figure 10.4   Evaluating τ

To see this, refine Fig. 10.2(ii) by dropping perpendiculars from D as shown in Fig. 10.3(ii); we want to show that θ, ϕ in Fig. 10.3(ii) are complementary. First evaluate τ, as follows. Draw two more diagonals DA, DB in Fig. 10.2(i), with DA meeting BE in F (see Fig. 10.4). Observe that $\triangle$s ABD, EFD are both isosceles and have the same base angle $2\pi/5$. They are therefore similar and so have proportional sides: $\tau/1 = 1/(\tau - 1)$, that is,

$$\tau^2 = \tau + 1, \tag{10.7}$$

and τ is none other than the **golden section**, $\tau = (1 + \sqrt{5})/2$.

Now, from Fig. 10.3(ii),

$$\cos\theta = \frac{\tau - 1}{2x}, \quad \cos\phi = \frac{\tau}{2y},$$

and by Pythagoras in Fig. 10.3(i),

$$x^2 = 1 - \frac{\tau^2}{4} = \frac{3 - \tau}{4}, \quad y^2 = 1 - \frac{(\tau - 1)^2}{4} = \frac{2 + \tau}{4},$$

whence

$$\begin{aligned}
\cos^2\theta + \cos^2\phi &= \frac{2 - \tau}{3 - \tau} + \frac{\tau + 1}{2 + \tau} \\
&= \frac{4 - \tau^2 + 3 + 2\tau - \tau^2}{6 + \tau - \tau^2} \\
&= \frac{5}{5} = 1.
\end{aligned}$$

Thus, $\cos^2\phi = \sin^2\theta$. Since θ and ϕ are both acute, we get $\cos\phi = \sin\theta$, so that $\theta + \phi = \pi/2$, as required.

This completes the construction of the five regular solids in $\mathbb{R}^3$. The interested or sceptical reader can verify this by cutting out little pieces of cardboard and sticking them together, or by calling at a shop in Kyoto, where you can actually buy them. Before embarking on our final task of this chapter, the calculation of symmetry groups, we make two remarks on **D** and **I**, leaving their proofs to the exercises. First, **D** and **I** are dual to one another, just like **C** and **O**. Second, they can also be described by specifying the coordinates of their vertices, and these can be expressed in a natural way in terms of the golden section (Exercises 10.11 and 10.12).

10.5 Symmetry Groups

Every symmetry of each of the five solids must fix the centre, and thus is either a rotation (OP) or a reflection (OR). Again in each case, the rotation subgroup Sym$^+$ has index 2 in the full symmetry group Sym, and multiplication by any reflection defines a one-to-one correspondence between Sym$^+$ and Sym$^-$ = Sym $\setminus$ Sym$^+$. Since dual pairs have the same symmetries, it is sufficient to compute Sym$^+$ and Sym for each of **T**, **C** and **D**. The results that follow were foreshadowed in Table 10.1, where the last column gives $g = |$Sym$^+|$ in each case.

First for **T**, rotations through $\pm 2\pi/3$ about each of the four axes joining the centre to a vertex permute the vertices $\mathcal{P}$ in all possible (eight) 3-cycles. Also, rotation through π about each of the three axes joining the centre with the midpoints of opposite edges produces the three permutations of shape $(\cdot\cdot)(\cdot\cdot)$ of $\mathcal{P}$. Throwing in the identity, we get $\mathrm{Sym}^+(\mathbf{T}) \cong A_4$. Since reflection in the plane containing the centre and two vertices transposes the other two, it follows that $\mathrm{Sym}(\mathbf{T}) \cong S_4$.

The cube $\mathbf{C}$ has four "grand diagonals" joining pairs of opposite vertices (and forming diameters of the circumscribing $\mathbb{S}^2$). The rotations of $\mathbf{C}$ act on this set as permutations of the following shapes:

- through $\pm\pi/2$ about three axes joining centres of opposite faces, $(\cdot\cdot\cdot\cdot)$;

- through π about the same axes, $(\cdot\cdot)(\cdot\cdot)$;

- through π about six axes joining midpoints of opposite edges, $(\cdot\cdot)$;

- through $\pm\pi/3$ about four axes joining opposite vertices, $(\cdot\cdot\cdot)$.

Throwing in the identity, we get $1 + 2 \cdot 3 + 3 + 6 + 2 \cdot 4 = 4!$, and $\mathrm{Sym}^+(\mathbf{C}) \cong S_4$. Reflection in the centre fixes each of the grand diagonals, reversing their orientations. It can be shown (Exercise 10.13) that $\mathrm{Sym}(\mathbf{C}) \cong S_4 \times Z_2$.

Turning to the dodecahedron, we seek a set of five objects on which the symmetries of $\mathbf{D}$ act as permutations from S_5. Consider the cube $\mathbf{C}$ of side τ from which $\mathbf{D}$ was constructed above. Each of the 12 edges of $\mathbf{C}$ is a diagonal in each of the twelve faces of $\mathbf{D}$. In accordance with Fig. 10.2(ii), $\mathbf{C}$ is determined by the diagonal appearing in any one face of $\mathbf{D}$. The other four diagonals of this face likewise determine four more cubes inscribed in $\mathbf{D}$, and all five cubes, like the diagonals determining them, are permuted cyclically by rotation s through $2\pi/5$ about the line joining O and the centre of this face. The cubes are also permuted by rotation s' through $2\pi/3$ about the line joining O to a vertex of $\mathbf{D}$. Since $s'^3 = 1$, this permutation is either the identity or a 3-cycle. Taking the axis of s' to be OD, with D as in Fig. 10.2(ii), the edge EE' of $\mathbf{C}$ passes under s' to $E'C$, which is not an edge of $\mathbf{C}$. So s' does not fix $\mathbf{C}$, and therefore acts as a 3-cycle.

It remains to determine the effect of rotation through π about the line joining O to the midpoint of an edge of $\mathbf{D}$. To do this, take a face $ABCDE$ of $\mathbf{D}$ with centre P and M the midpoint of CD, as shown in Fig. 10.5, and let r_1, r_2, r_3 denote reflection in the planes POM, COP, COM, respectively. If s, s' are clockwise rotations about OP, OC through $2\pi/5$, $2\pi/3$, respectively, then by Theorem 4.1,

$$s = r_1 r_2, \quad s' = r_2 r_3,$$

whence

$$ss' = r_1 r_3$$

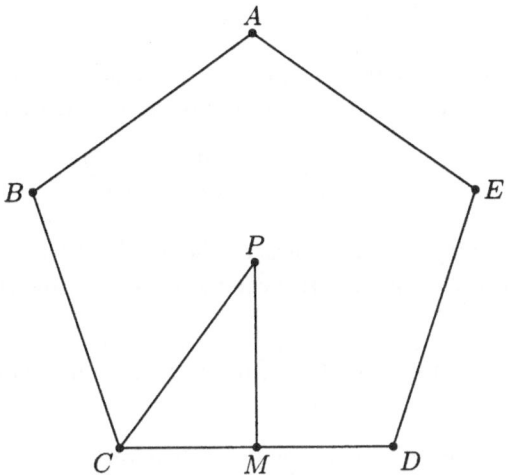

Figure 10.5   Counting rotations

is rotation through π about OM. Since this permutation is not the identity it has order two, and since it is even it has the shape $(\cdot\cdot)(\cdot\cdot)$. Thus, the rotational symmetries of **D** are

$6 \cdot 4$ of type s with shape $(\cdot\cdot\cdot\cdot\cdot)$,
$10 \cdot 2$ of type s' with shape $(\cdot\cdot\cdot)$,
$15 \cdot 1$ of type ss' with shape $(\cdot\cdot)(\cdot\cdot)$.

Throwing in the identity, we get $1+6.4+10.2+15.1 = 5!/2$, and $\mathrm{Sym}^+(\mathbf{D}) \cong A_5$. Extending this group by reflection in O, we get (Exercise 10.14) $\mathrm{Sym}(\mathbf{D}) = A_5 \times Z_2$.

EXERCISES

10.1. What word would you use to describe a sphere in $\mathbb{R}^2$ with (a) the Pythagorean, (b) the Manhattan metric?

10.2. Say which of the following facts in plane geometry remain valid in the spherical case and which do not. Give reasons.

(a) An exterior angle of a triangle is equal to the sum of the interior opposites.

(b) Two triangles are congruent if two sides and the included angle are equal.

(c) The diagonals of a parallelogram bisect one another.

(d) The theorem of Pythagoras.

10.3. Prove the **sine rule** for spherical triangles: if a triangle has angles α, β, γ and opposite sides a, b, c respectively, then

$$\frac{\sin \alpha}{\sin a} = \frac{\sin \beta}{\sin b} = \frac{\sin \gamma}{\sin c}.$$

10.4. Let L be a lune with vertices at the north and south poles and angle α. By integrating with respect to latitude, show that the area of L is 2α.

10.5. Prove that any triangle PQR on $\mathbb{S}^2$ is congruent to its antipode $P'Q'R'$.

10.6. Let Γ be a graph in the plane with v vertices, e edges and f faces (suitably defined). Use Theorem 10.3 to prove that

$$v - e + f = c,$$

the number of connected components of Γ.

10.7. Prove that the number of n-gons in a regular (m, n)-tessellation of $\mathbb{S}^2$ is equal to $\frac{4m}{4-(m-2)(n-2)}$.

10.8. Describe and compare the degenerate regular tessellations $(2, m)$ and $(m, 2)$ of $\mathbb{S}^2$. Compute their parameters v, e, f and identify their symmetry groups.

10.9. Letting $\theta = \pi/10$, use the fact that $\cos 3\theta = \sin 2\theta$ to calculate $\sin \theta$. Deduce that the golden section

$$\tau = 2 \cos \pi/5 = \frac{1 + \sqrt{5}}{2}.$$

10.10. Use Table 10.1 to show that **D** and **I** are dual to one another.

10.11. Put $t = \sqrt{\tau}$, where τ is the golden section, and consider the set V of 12 points

$$(\pm t, \pm t^{-1}, 0), \quad (0, \pm t, \pm t^{-1}), \quad (\pm t^{-1}, 0, \pm t)$$

in $\mathbb{R}^3$. By calculating the distances between various pairs of them and using symmetry, prove that V is the vertex set of a regular icosahedron on the sphere centre O radius $\sqrt[4]{5}$.

10.12. Use duality and the previous exercise to find the coordinates of the 20 vertices of a regular dodecahedron on $\mathbb{S}^2$.

10.13. Prove that $\mathrm{Sym}(\mathbf{C}) \cong S_4 \times Z_2$.

10.14. Prove that $\mathrm{Sym}(\mathbf{D}) \cong A_5 \times Z_2$.

10.15. Show that the volumes of the five regular solids in $\mathbb{S}^2$ are, in some order,

$$8\sqrt{3}/9, \ 4/3, \ 4\sqrt{3}/9, \ 4\sqrt{3}(\tau+2)/9, \ 4\sqrt{\tau+2}/9,$$

where τ is the golden section.

10.16. Specify the angle between a pair of adjacent faces in each of the five regular solids.

10.17. Use the result of the previous exercise to show that there is one and only one regular tessellation of $\mathbb{R}^3$.

11
Triangle Groups

Let x, y, z denote reflections in the three sides of a triangle. Then the product of any two of these is rotation about a vertex through twice the interior angle. Thus, for the group Δ generated by x, y, z to be discrete, a necessary condition is that each interior angle be an integer submultiple of π, say π/l, π/m, π/n, where the integers l, m, n are

(a) all at least 2, to avoid degeneracy, and

(b) enumerated in such a way that $l \leq m \leq n$.

We then have relations

$$x^2 = y^2 = z^2 = 1 = (xy)^l = (yz)^m = (zx)^n, \tag{11.1}$$

which define the **triangle group** $\Delta(l, m, n)$ on the generators x, y, z.

It is clear that the elements $a = xy$, $b = yz$ are OP and satisfy the relations

$$a^l = b^m = (ab)^n = 1. \tag{11.2}$$

It is left to the exercises to show that the OP subgroup $\Delta^+(l, m, n)$ is generated by a, b and defined by the relations (11.2).

Applying the elements of $\Delta(l, m, n)$, we obtain a tessellation of the ambient space by copies of the original triangle. But what is this space? The answer, as outlined below, depends critically but solely on the angle-sum of the triangle. There are three cases, of which the first two have already been studied.

11.1 The Euclidean Case

Since the angle-sum of a triangle in $\mathbb{R}^2$ is equal to π, we have to solve the Diophantine equation

$$1/l + 1/m + 1/n = 1.$$

Under the conditions $2 \leq l \leq m \leq n$, there are just three solutions,

$$(l, m, n) = (2, 3, 6), \ (2, 4, 4), \ (3, 3, 3), \tag{11.3}$$

and the corresponding triangles are the familiar ones depicted in Fig. 11.1. We describe the tessellations and groups corresponding to each in turn.

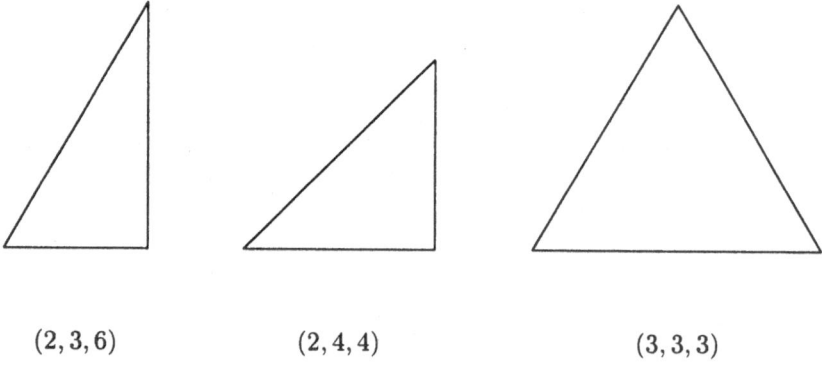

$(2, 3, 6)$ $(2, 4, 4)$ $(3, 3, 3)$

Figure 11.1   Euclidean values of (l, m, n)

Consider the regular tessellation $(3, 6)$ of $\mathbb{R}^2$ by hexagons meeting in threes (see Fig. 9.2). Let F be one of the hexagonal tiles and T the triangle OPQ, where O is the centre of F, P is a vertex of the boundary, and Q is the midpoint of an edge incident to P. Then T is a $(2, 3, 6)$ triangle and its 12 images under $\mathrm{Sym}(F) \cong D_{12}$ cover F. The required tessellation is then obtained from $(3, 6)$ by similarly inserting 12 "spokes" in every tile. This process is called **barycentric subdivision** and forms the basis of a geometric proof that the corresponding triangle group is isomorphic to the symmetry group of $(3, 6)$,

$$\Delta(2, 3, 6) \cong \mathrm{Sym}(3, 6) = G_6^1. \tag{11.4}$$

Because of duality, the application of this construction to the tessellation $(6, 3)$ in Fig. 9.3 leads to the same result.

We next apply this construction to the tessellation $(4, 4)$ in Fig. 9.1, so that each of the square tiles is partitioned into eight copies of a $(2, 4, 4)$ triangle. This is the required tessellation, and we see that

$$\Delta(2, 4, 4) \cong \mathrm{Sym}(4, 4) = G_4^1. \tag{11.5}$$

Since there are no other regular tessellations of $\mathbb{R}^2$, the tessellation by $(3,3,3)$ triangles cannot be obtained by the above construction. We leave as an exercise the intriguing question of whether or not $\Delta(3,3,3)$ is isomorphic to a plane crystallographic group.

A general remark needs to be made at this point concerning the isomorphisms in (11.4) and (11.5). While they certainly hold (see Exercises 11.5 and 11.6), the astute reader will have noticed a certain lack of rigour in the geometric proof. To spell this out, take the case of $\Delta(2,3,6)$ acting as described on the corresponding tessellation. What we have not shown is that this action is **faithful**, that is, if an element $w = w(x,y,z) \in \Delta(2,3,6)$ fixes the tessellation tilewise, then $w = 1$ in the group. (In contrast to this, G_6^1 acts faithfully by definition.) The following sketch of a proof provides an illustration of the beginnings of a subject that has come to be called "geometric group theory."

Let $\mathcal{T}$ be the tessellation of $\mathbb{R}^2$ by $(2,3,6)$ triangles as described above, and let $\mathcal{T}^*$ be the **dual tessellation** obtained from $\mathcal{T}$ as follows. For each triangle in $\mathcal{T}$ take a point in its interior, say its in-centre, the point of concurrence of its angle-bisectors: these points are the vertices of $\mathcal{T}^*$. Two such points are joined by an edge in $\mathcal{T}^*$ if and only if the corresponding triangles in $\mathcal{T}$ abut along an edge. There is then one tile of $\mathcal{T}^*$ for each vertex of $\mathcal{T}$, and these tiles are squares, hexagons or dodecagons according as the corresponding vertex in $\mathcal{T}$ has valency 4, 6 or 12. The edges of $\mathcal{T}^*$ are then labelled x, y or z in such a way that the boundary labels of these polygons are equal to $(xy)^2$, $(yz)^3$ or $(zx)^6$ according to case and the three edges of $\mathcal{T}^*$ incident to any vertex have distinct labels. Words in the alphabet x, y, z thus correspond to edge-paths in $\mathcal{T}^*$, and there is a unique such path with a given starting point and labelled by a given word.

We now suppose that $w = w(x,y,z)$ is a word that fixes just one tile in $\mathcal{T}$ and claim that $w = 1$ in $\Delta(2,3,6)$. Our assumption means that w fixes a vertex in $\mathcal{T}^*$, which implies that the corresponding edge-path is a circuit. This circuit along with its interior thus forms a van Kampen diagram (see Fig. 5.1). This means that the boundary label w of the diagram is a product of conjugates of the boundary labels of its cells (the proof is by induction on the number of cells). Since the latter are all relators, we have $w = 1$ in $\Delta(l,m,n)$, as required.

The above sketch can be made rigorous and extended to prove the following general result.

Theorem 11.1

The action described above of the group $\Delta(l,m,n)$ on a tessellation $\mathcal{T}$ by (l,m,n) triangles is without fixed points. □

Since this action is obviously **transitive** (given any two tiles in $\mathcal{T}$ there is
an element of $\Delta(l, m, n)$ sending one to the other), it follows that the action is
actually **regular**, that is, given any two tiles there is exactly one such element.
The tiles in $\mathcal{T}$ are thus in one-to-one correspondence with the elements of the
group, such a correspondence being determined by nominating one particular
tile, called a **fundamental region**, to correspond to the identity 1 of the group.

It follows that the elements of the group are also in one-to-one correspondence with the vertices of the labelled graph $\mathcal{T}^*$, which (together with a distinguished vertex) is called the **Cayley diagram** of the group with respect to the
given generators. Note that this concept is quite general: to define the Cayley
diagram of a group G with respect to a finite set X of generators, we merely
need to attach an orientation (arrow) to each edge labelled by a generator x of
order $\neq 2$. Traversing such an edge with or against the arrow then corresponds
to multiplication by x or x^{-1}, respectively.

11.2 The Elliptic Case

To solve the Diophantine inequality

$$1/l + 1/m + 1/n > 1$$

subject to the conditions $2 \leq l \leq m \leq n$, first notice that we must have $l = 2$.
Then we can take $m = 2$ and any value of $n \geq 2$, or $m = 3$ and $n = 3$, 4 or 5.
Since we cannot have $m \geq 4$, the list is complete:

$$(l, m, n) = (2, 2, n), \ (2, 3, 3), \ (2, 3, 4), \ (2, 3, 5), \tag{11.6}$$

where $n \geq 2$. Since we want a space in which the angle-sum of a triangle exceeds
π, the strong resemblance of this calculation to the last step in the proof of
Theorem 10.4 is no accident: the solutions (11.5) all parametrise tessellations
of the sphere $\mathbb{S}^2$.

First take the infinite family $(2, 2, n)$, $n \geq 2$. The Greenwich meridian in-
tersects the equator at right angles, and so does any other line (semicircle) of
longitude such as $180/n°$E. These three lines clearly bound a $(2, 2, n)$ triangle
T in the **N** hemisphere, which is partitioned into $2n$ congruent triangles by
reflecting alternately in the two longitudinal sides. Reflection in the equatorial
side (or in the centre of the sphere) yields a similar partition of the **S** hemisphere, and we obtain a tessellation $\mathcal{T}$ of $\mathbb{S}^2$ by $4n$ congruent $(2, 2, n)$ triangles.
The relations (11.2) define a dihedral group, and we have

$$\Delta^+(2, 2, n) \cong D_{2n}.$$

And since reflection y in the equator (or centre) is central (it commutes with x and z by (11.1) with $l = m = 2$), we get

$$\Delta(2, 2, n) \cong D_{2n} \times Z_2.$$

The fact that the order of this group and the number of tiles in $\mathcal{T}$ are equal (to $4n$) provides an illustration of Theorem 11.1. The triangle T is a fundamental region, and the Cayley diagram of the group, viewed from the **N** pole, is shown in Fig. 11.2 in the case $n = 3$.

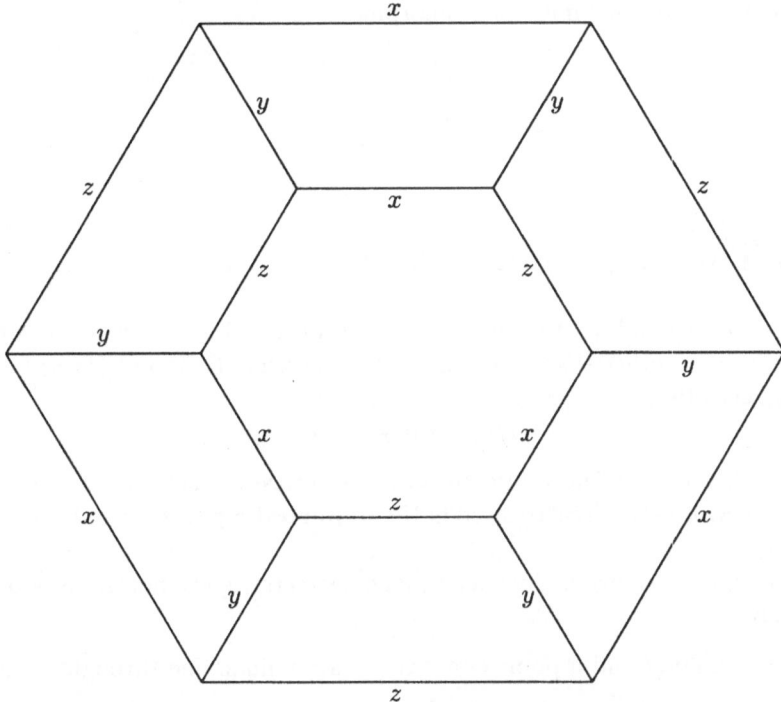

Figure 11.2   Cayley diagram for $\Delta(2, 2, 3)$

To realise the solutions $(2, 3, n)$, $n = 3, 4, 5$, consider the regular tessellations $(n, 3)$ of $\mathbb{S}^2$ defined in Chapter 10 (see Table 10.1) for the same values of n. The tiles in the latter are equilateral triangles with angle $2\pi/n$, and the spoke construction of the previous section partitions each of them into six congruent $(2, 3, n)$ triangles. This is the required tessellation in each of the three cases. The number of triangles is $6f = 2g$, where f and g are as in rows **T**, **O**, **I** of Table 10.1, namely 24, 48, 120 when $n = 3, 4, 5$ respectively. By Theorem 11.1, these numbers are the orders of the corresponding triangle groups, which we now identify as symmetry groups in accordance with the results of Chapter 10:

$$\Delta^+(2,3,3) \cong A_4, \quad \Delta^+(2,3,4) \cong S_4, \quad \Delta^+(2,3,5) \cong A_5,$$
$$\Delta(2,3,3) \cong S_4, \quad \Delta(2,3,4) \cong S_4 \times Z_2, \quad \Delta(2,3,5) \cong A_5 \times Z_2.$$

An alternative derivation of these isomorphisms is as follows. In the OP case, see (5.16) and (5.18) for $n = 3, 4$, and for $n = 5$, consider the permutations $\alpha = (12)(34)$, $\beta = (245)$. It is easy to show that they generate the alternating group A_5 and satisfy the relations $\alpha^2 = \beta^3 = (\alpha\beta)^5 = 1$, and so A_5 is a homomorphic image of $\Delta^+(2,3,5)$. But then, since the orders of these groups are equal (to 60), they must be isomorphic:

$$\Delta^+(2,3,5) \cong A_5.$$

For the general case, see Exercise 11.3.

11.3 The Hyperbolic Case

It follows from the law of trichotomy in $\mathbb{R}$ that all triples (l, m, n) of integers, $2 \le l \le m \le n$, other than those in the (short) lists (11.3) and (11.6) fall into the **hyperbolic case**
$$1/l + 1/m + 1/n < 1,$$

which is therefore the most numerous by a street. Nevertheless, the corresponding tessellations can all be realised in the **hyperbolic plane** $\mathbb{H}^2$, which we now describe.

Among the axioms of Euclidean plane geometry is the notorious Axiom of Parallels,

AP: given a line l and a point $O \not\in l$ there is a unique line through O parallel to l.

The question that occupied many mathematicians for many years is whether or not this axiom is independent of the others (axioms of incidence, order, congruence and continuity). The question was finally settled by Bolyai, Gauss and Lobachevskii more or less independently. The answer "yes, it is independent" is established by the existence of the hyperbolic plane $\mathbb{H}^2$, famous models of which include those of Klein and Poincaré. The most convenient for our purposes is the **Poincaré disc model**, which looks like this.

Let D be an open disc in the Euclidean plane. The word "open" here means that we exclude all points of the circular boundary ∂D of D; the points of ∂D are called the **ideal points** of the model. Through any two ideal points there is a unique line (if they are diametrically opposed) or circle orthogonal to ∂D (if they are not). Then the segments of these lines or circles that lie in D are the **lines** of $\mathbb{H}^2$, the **points** being just those in D. It turns out that all Euclid's axioms except AP hold in this space.

Let us show that $\mathbb{H}^2$, in the form of the Poincaré disc model just described, contains a copy of any hyperbolic triangle, that is, any triangle with angle-sum less than π. While there is a shorter, if rather less satisfactory, proof using the Intermediate Value Theorem from Real Analysis, we prefer the following constructive approach.

Let α, β, γ, δ be positive real numbers such that $\alpha + \beta + \gamma + \delta = \pi$. Then there is a Euclidean triangle ABC with

$$\widehat{A} = \alpha, \quad \widehat{B} = \beta + \delta/2, \quad \widehat{C} = \gamma + \delta/2.$$

Next let P be the interior point of $\triangle ABC$ which is the apex of an isosceles triangle PBC with base angle

$$P\widehat{B}C = P\widehat{C}B = \delta/2.$$

Then there is a unique circle C_1 tangent to the lines PB, PC at the points B, C respectively, and a unique circle C_2 centre A orthogonal to C_1. Taking the disc D to be the interior of C_2, the hyperbolic triangle ABC is the one we want: it is obtained from its Euclidean namesake by replacing the straight line segment BC by the (clockwise) arc BC of the circle C_1.

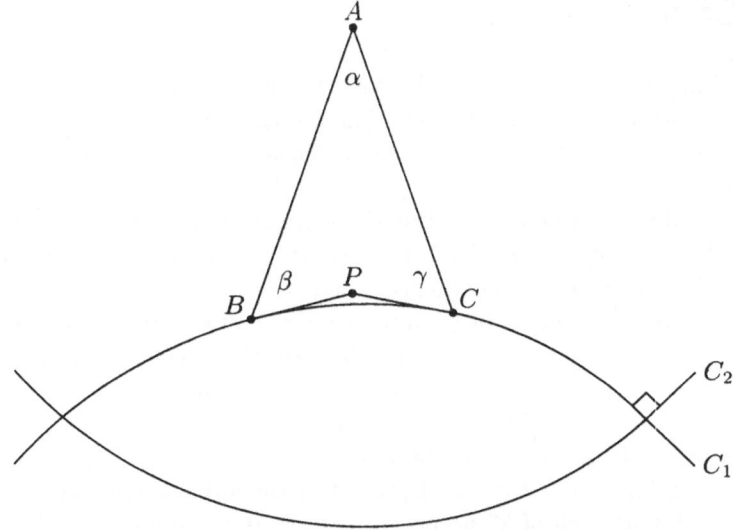

Figure 11.3   Construction of a hyperbolic triangle

Theorem 11.2

For any integers l, m, n such that $2 \leq l \leq m \leq n$ and $1/l + 1/m + 1/n < 1$, there is a triangle in $\mathbb{H}^2$ with angles π/l, π/m, π/n. □

The above construction clearly proves more than this; the angles α, β, γ need not be reciprocals of integers. Indeed, any or all of them can be zero, in which case the corresponding vertices will be ideal points. More generally, the construction is not limited to triangles. For example, it is easy to see that $\mathbb{H}^2$ contains regular ideal n-gons for all $n \geq 3$. Thus the structure of $\mathbb{H}^2$ is arguably richer than that of $\mathbb{R}^2$, and the same holds for the corresponding triangle groups.

Table 11.1   The three geometries

	$\mathbb{S}^2$	$\mathbb{R}^2$	$\mathbb{H}^2$
Angle-sum of Δ	$> \pi$	$= \pi$	$< \pi$
Parallels	None	Unique	Many
Curvature	Positive	Zero	Negative
Triangle groups G	G finite	$|G : Z^2|$ finite	G contains F_2

11.4 Coxeter Groups

The remaining sections of this book represent an attempt to escape from the plane, of whatever kind, into higher dimensions. The spaces $\mathbb{R}^n$, $\mathbb{P}^n$, $\mathbb{H}^n$ all have models of the kind described above in the case $n = 2$. We say a few words about each.

Most familiar is perhaps real Euclidean n-space $\mathbb{R}^n$, which, when $n = 3$, provides the Newtonian model of the cosmos. The objects corresponding to polygons when $n = 2$ and polyhedra when $n = 3$ are called **polytopes** in general. Pursuing the theme of Chapter 10, we shall classify the regular polytopes for all n in the next chapter: the list contains some pleasant surprises. The crystallographic groups when $n = 3$ are of particular interest to chemists; there are 230 altogether. The corresponding list for $n = 4$ is also finite and is contained in an electronic file at the RWTH in Aachen. The fact that the list is finite for any fixed $n \geq 2$ was first proved by Bieberbach and solves a famous problem of Hilbert, one of 23 he posed in 1900 at the International Congress of Mathematicians in Paris.

The symbol $\mathbb{P}^n$ denotes n-dimensional projective space. Note that the real projective plane $\mathbb{P}^2$ is the preferred model for elliptic space in dimension two. It is formed by "identifying" pairs of diametrically opposed points of the sphere $\mathbb{S}^2$. Here the group $G = \{1, r\} \cong Z_2$ acts on $\mathbb{S}^2$ by letting 1 be the identity map and r the reflection in the centre. Then $\mathbb{P}^2$ is the resulting quotient space, to be thought of in the same way as a quotient group (or factor group) as defined in Chapter 3. The discrete groups of isometries of elliptic space are finite in all dimensions.

The preferred model for $\mathbb{H}^n$, also due to Poincaré, is the upper half-space $\{(x_1, \ldots, x_n) \in \mathbb{R}^n \mid x_n > 0\}$. Its boundary is a copy of $\mathbb{R}^{n-1}$ and arcs of circles (and segments of lines) orthogonal to this boundary comprise the lines of $\mathbb{H}^n$. The study of this space, which involves a beautiful blend of geometry, group theory and analysis, is sadly beyond our scope. We make only a few sketchy remarks about regular polyhedra and tessellations in the ball model of $\mathbb{H}^3$. For the algebraist at least, these are best understood in the light of the following definition.

Definition 11.1

A **Coxeter group** G is one defined by a presentation of the following kind:

generators $x_1, x_2, \ldots, x_n,$

relators $x_i^2, \ 1 \leq i \leq n, \ (x_i x_j)^{m_{ij}}, \ 1 \leq i < j \leq n.$

Here n is a positive integer called the **rank** of G and the integer parameters m_{ij} are its **exponents**. The latter are assumed to be ≥ 0 (without loss of generality) but $\neq 1$ (to avoid degeneracy).

In rank 1 there are no exponents and we trivially have $G \cong Z_2$. When $n = 2$ G is a dihedral group, of order $2m_{12}$ when $m_{12} \in \mathbb{N}$ and of infinite order when $m_{12} = 0$. In the case $n = 3$ we get the triangle groups studied above, of which Coxeter groups are an obvious formal generalisation. We give an example of rank 4 below.

The notion of reflection in a line in $\mathbb{R}^2$ can be extended in various ways. For example, we may allow

(a) orthogonal reflection in a hyperplane in $\mathbb{R}^n$,

(b) inversion of a circle (or sphere) as in $\mathbb{H}^2$; or

(c) linear transformations of $\mathbb{R}^n$ that fix pointwise a hyperplane through the origin and map some non-zero point to minus itself.

It is in this last sense that the alternative name "groups generated by reflections" is applied to Coxeter groups. For then any Coxeter group G of rank n acts faithfully on $\mathbb{R}^n$, so that G can be embedded as a subgroup in the group $GL(n, \mathbb{R})$ of all non-singular $n \times n$ matrices over $\mathbb{R}$.

Coxeter groups can be classified in a manner similar to that summarised for triangle groups in Table 11.1. The finite ones are known, and so are those that contain a free abelian subgroup of finite index (these are called **affine**): see Figs. 11.5 and 11.6 at the end of this section. The term "hyperbolic" has several meanings, both geometric and group theoretic, when applied to Coxeter groups. The example given below of a Coxeter group acting on $\mathbb{H}^3$ is of this kind. To get an overall picture, one needs a general type of structure on which any Coxeter group can be made to act in a natural way. Such a concept was provided by J. Tits in his theory of buildings.

Coxeter groups are very numerous: $n(n-1)/2$ parameters are required in rank n. These can be specified in the form of a matrix $M = (m_{ij})$, where $m_{ii} = 1$ for all i and $m_{ij} = m_{ji}$ for $i > j$. A more popular and visual method is to use a **Dynkin diagram** (or **Coxeter graph**). This is a graph with n vertices, one for each generator, in rank n, and edges as follows:

(a) no edge joining i and j when $m_{ij} = 2$,

(b) an unlabelled edge from i to j when $m_{ij} = 3$,

(c) an edge labelled m_{ij} from i to j when $m_{ij} \geq 4$,

(d) an edge labelled ∞ from i to j when $m_{ij} = 0$,

so that the label stands for the order of the element $x_i x_j$. For example, the complete graphs on 2, 3 and 4 vertices (see Fig. 11.4) stand for the dihedral group D_6, the triangle group $\Delta(3,3,3)$ and the following hyperbolic Coxeter group.

Example 11.1

Take a regular tetrahedron inscribed in a sphere $\mathbb{S}^2$ and replace each of its four plane triangular faces by the corresponding part of the sphere through its three vertices and orthogonal to $\mathbb{S}^2$. The result is an ideal regular hyperbolic tetrahedron. The angles between adjacent edges are all zero and it is an exercise to show that the angles between adjacent faces are all equal to $\pi/3$. This means that the reflections in the four faces generate the Coxeter group with Dynkin diagram K_4; see Fig. 11.4(iii).

We end this chapter by giving the classification of finite and affine Coxeter groups. The irreducible ones are listed in the form of Dynkin diagrams in Figs.

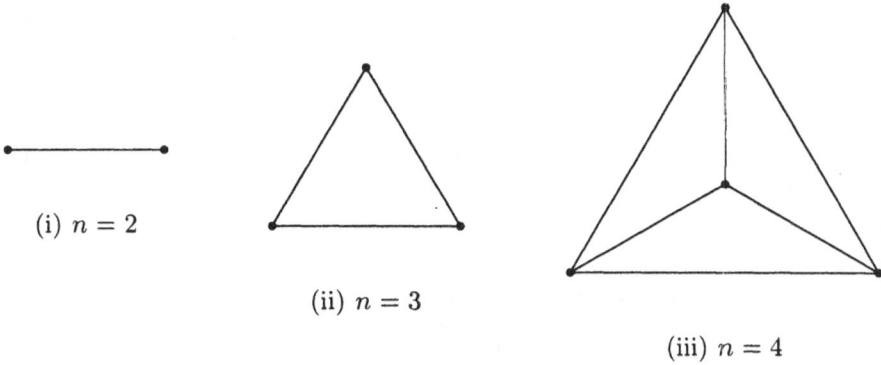

(i) $n = 2$

(ii) $n = 3$

(iii) $n = 4$

Figure 11.4   The complete graph K_n

11.5 and 11.6. (Here a Coxeter group G is **irreducible** if there is no partition of the generating set into two subsets A and B such that every a in A commutes with every $b \in B$. If there is such a partition, then $G \cong \langle A \rangle \times \langle B \rangle$ and its Dynkin diagram is disconnectd.) The proof, like that of the analogous result for simple Lie algebras of finite dimension over $\mathbb{C}$, is a tour de force. Most of the groups in the finite case (Fig. 11.5) will appear as symmetry groups in the next chapter. The suffices indicate the rank of the group in Fig. 11.5 and rank minus 1 in Fig. 11.6.

EXERCISES

11.1. Use Tietze transformations informally to establish the isomorphisms

$$\Delta(l, m, n) \cong \Delta(m, l, n) \cong \Delta(m, n, l).$$

Deduce that the group $\Delta(l, m, n)$ depends only on the set $\{l, m, n\}$ and not on the triple (l, m, n).

11.2. Let G be the triangle group defined by the relations (11.1) and H the subgroup generated by $a = xy$, $b = yz$. Use the first three relations in (11.1) to show that H contains every word of even length in x, y, z and deduce that $|G : H| \leq 2$. Obtain the reverse inequality by studying the factor group G/H.

11.3. Let H be the group defined by the relations (11.2) and check that the map $a \mapsto a^{-1}$, $b \mapsto b^{-1}$ defines an automorphism α of H. Show that the resulting semidirect product $Z_2 \times_\alpha H$ has a presentation equivalent to that defining $\Delta(l, m, n)$.

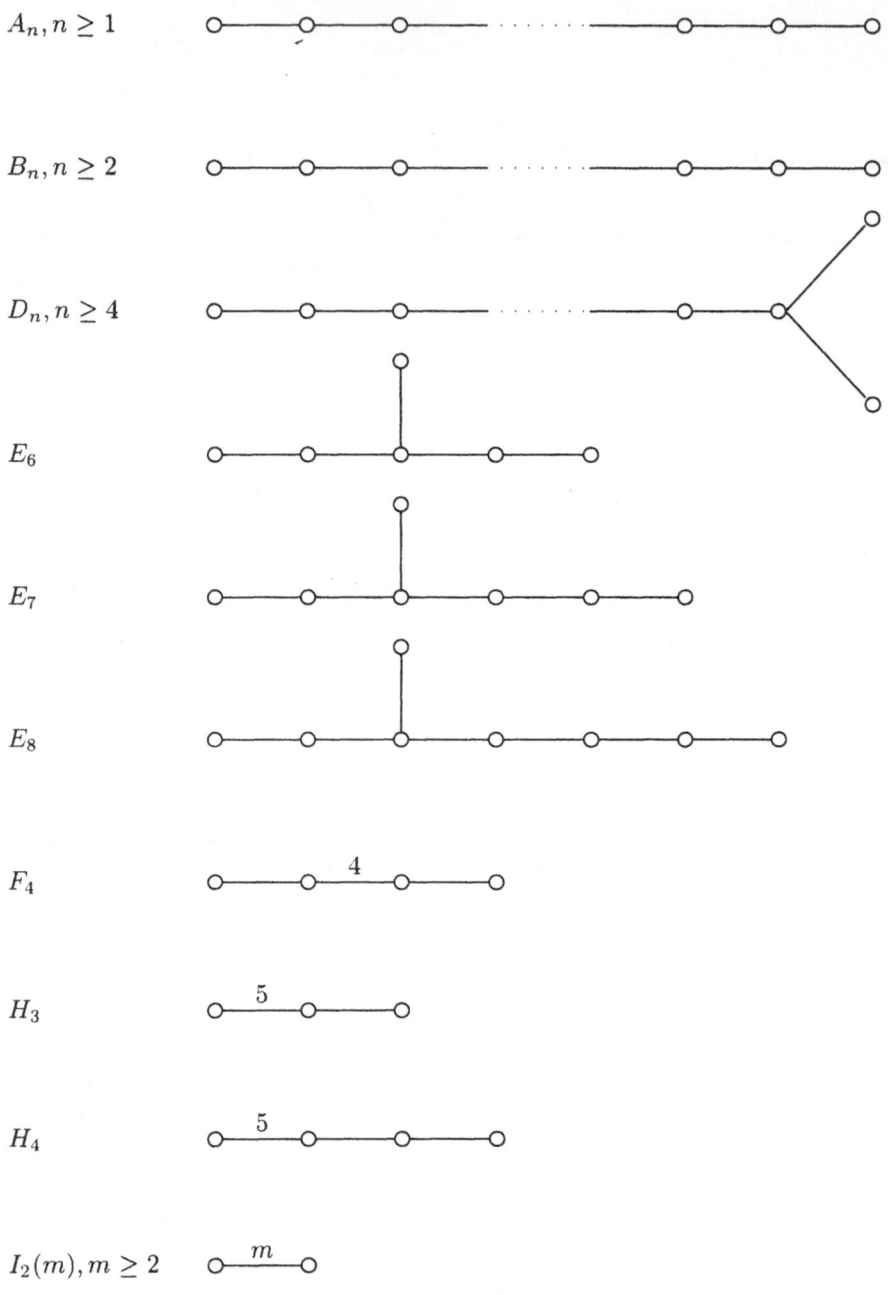

Figure 11.5   Finite Coxeter groups

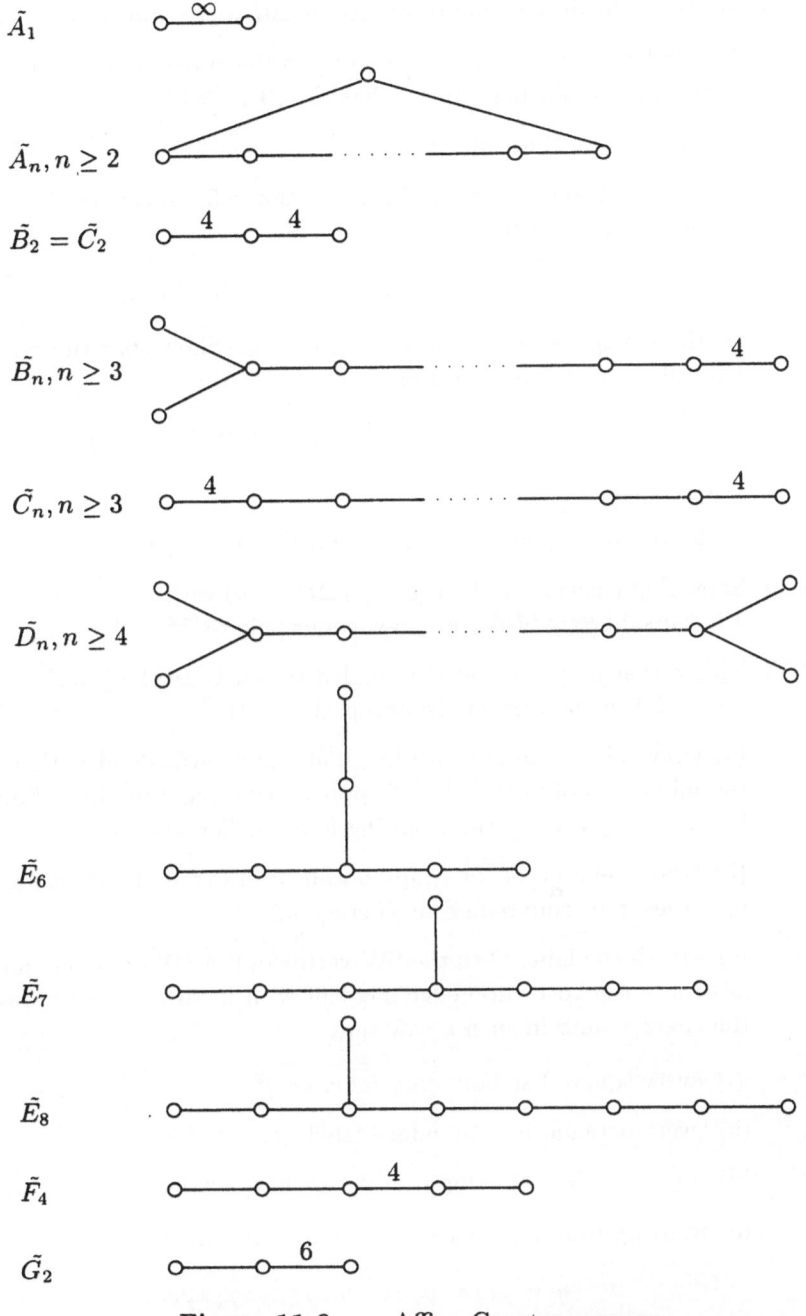

Figure 11.6   Affine Coxeter groups

11.4. Describe the derived factor groups of $\Delta(l, m, n)$ and $\Delta^+(l, m, n)$.

11.5. By applying Tietze transformations to the corresponding presentations, give an algebraic proof that $\Delta(2, 3, 6) \cong G_6^1$.

11.6. Likewise prove that $\Delta(2, 4, 4) \cong G_4^1$.

11.7. Use Tietze transformations (as in Section 5.3, especially (5.14)), to obtain a presentation of the group

$$\Delta^+(3, 3, 3) = \langle s, t \mid s^3 = t^3 = (st)^3 = 1\rangle$$

on the new generators s, $a = ts^{-1}$, $b = tst$. Show that the result is equivalent to the presentation

$$\langle a, b, s \mid ab = ba, s^3 = 1, a^s = a^{-1}b, b^s = a^{-1}\rangle$$

of G_3 in Theorem 7.1.

11.8. Is the triangle group $\Delta(3, 3, 3)$ isomorphic to G_3^1 or G_3^2?

11.9. Show that each of the three groups $\Delta(l, m, n)$ with $1/l + 1/m + 1/n = 1$ has a subgroup of finite index isomorphic to Z^2.

11.10. This rather long exercise is intended to illustrate the proof of Theorem 11.1 in the case of the group $\Delta(2, 4, 4)$.

(a) Draw a large square with horizontal and vertical sides, then join the midpoints of both pairs of opposite sides, and finally subdivide barycentricly each of the resulting four smaller squares.

(b) Draw the dual of the graph obtained in (a): it should consist of nine tiles, four square and five octagonal.

(c) Attach the label O to the SW vertex of the SW octagon and the label y to the horizontal edge incident with it. Now label the rest of the edges x, y, z in such a way that

(i) every square has boundary label $(xy)^2$,

(ii) every octagon has boundary label $(yz)^4$ or $(zx)^4$,

(iii) any two edges meeting at a vertex have distinct labels.

(d) Starting from O, trace out the path with label

$$w = yzxz \cdot yzyz \cdot xyzy \cdot xyzy \cdot xzyz;$$

you should finish up back at O.

(e) Express w as a product of conjugates of the relators

$$r = (xy)^2, \quad s = (yz)^4, \quad t = (zx)^4$$

(which should appear four times, twice, once respectively), and deduce that $w = 1$ in $\Delta(2, 4, 4)$.

11.11. Show that, with respect to suitable sets of generators, the Cayley diagrams of Z_8 and D_8 can be drawn as regular octagons (with oriented edges in the former case).

11.12. Draw Cayley diagrams, in the plane if possible, for the other three groups of order 8.

11.13. What are the derived factor groups of the three Euclidean triangle groups? Show that in each case the derived group is free abelian of rank 2.

11.14. Use the Intermediate Value Theorem to supply a shorter, if rather less satisfactory, proof of Theorem 11.2.

11.15. Write down a presentation of the Coxeter group generated by reflections in the sides of an ideal equilateral hyperbolic triangle.

11.16. Can you identify the Coxeter group with diagram •——— • ———•? (Hint: it has order 24.)

11.17. Prove that the angles between adjacent faces of an ideal regular hyperbolic tetrahedron are all equal to $\pi/3$.

11.18. Say which of the 12 plane crystallographic groups in Chapter 8 are generated by reflections.

11.19. Identify the triangle groups with $1/l + 1/m + 1/n > 1, = 1$ in Figs. 11.5 and 11.6 respectively.

12
Regular Polytopes

We limit ourselves to a brief glimpse into this vast subject, confining attention to regular configurations in Euclidean space. The groups involved will be discrete and play a fundamental role: they arise even in the definition of the objects we wish to describe, which are natural generalisations to arbitrary dimension d of regular polygons in dimension 2 and polyhedra in dimension 3.

A **polytope** in $\mathbb{R}^d$ is an intersection of finitely many closed half-spaces that is bounded and has non-empty interior. A brief discussion of some of the terms involved in this definition may help to bolster intuition derived from the familiar cases $d = 2$ and $d = 3$. Thus, P is **bounded** if all its points lie within some given distance of a distinguished point $O \in \mathbb{R}^d$. **Subspaces** are translates (additive cosets) of their namesakes in linear algebra, and are called **lines**, **planes** and **hyperplanes** in dimensions 1, 2 and $d - 1$, respectively. It is intuitively reasonable and not hard to prove that a hyperplane H divides $\mathbb{R}^d$ into two "halves", called **half-spaces**. Such a half-space S is **closed** if it contains its boundary $H = \partial S$, the other points of S making up its **interior** $S^0 = S \setminus H$. The definition $P = \bigcap_{i=1}^n S_i$ is assumed to be irredundant (so that n is minimal), and we define $P^0 = \bigcap_{i=1}^n S_i^0$.

Let $P = \bigcap_{i=1}^n S_i$ as above with n minimal and put $H_i = \partial S_i$, $i = 1, 2, \ldots, n$. Then $P \cap H_i$ is called a **face** of P. For $k = d - 1, \ldots, 1, 0$, the k-**faces** of P are defined inductively as follows. A $(d-1)$-face is just a face of P. For $0 \le k < d-1$, a k-face is a face of a $(k + 1)$-face of P. A 1-face is called an **edge**, and a 0-face a **vertex**, of P. Two k-faces are said to be **adjacent** if their intersection is a $(k - 1)$-face. Notice that any edge contains exactly two vertices, and dually any $(d - 2)$-face is contained in exactly two faces, $P \cap H_i$ and $P \cap H_j$ say. Then the angle between (unit vectors orthogonal to) H_i and H_j is called a **dihedral angle** of P. A **flag** of P is a d-tuple $(F_0, F_1, \ldots, F_{d-1})$ of k-faces F_k such that $F_k \subseteq F_{k+1}$ for $k = 0, 1, \ldots, d-2$.

We can now give three equivalent definitions of regularity in application to a polytope P. P is **regular** if

(a) all its faces are isometric regular polytopes in dimension $d-1$ and all its dihedral angles are equal,

(b) all its faces are isometric regular polytopes in dimension $d-1$ and, for any vertex x of P, the other endpoints of edges of P containing x all lie in the same hyperplane H and comprise the vertex set of a regular polytope in H, or

(c) the group $G = \mathrm{Sym}(P)$ acts regularly on the flags of P.

The last condition asserts that, for any two flags $(F_0, F_1, \ldots, F_{d-1})$, $(F_0', F_1', \ldots, F_{d-1}')$ of P, there is exactly one $g \in G$ such that $F_i g = F_i'$, $0 \le i \le d-1$.

All this is rather elaborate. Even so there are gaps, some of which are filled by the exercises. Starting from the definition (c), we aim to classify all regular polytopes. The result is known already in dimensions $d = 2, 3$ and is fairly predictable when $d \ge 5$. The novelty comes in dimension 4.

12.1 The Standard Examples

There are three of these, and they exist in all dimensions $d \ge 2$. We begin with the easiest to picture.

In the usual parametrisation of Euclidean d-space by real d-tuples $\mathbf{x} = (x_1, \ldots, x_d)$, there are exactly 2^d points whose coordinates all have modulus 1. These are the vertices $(\pm 1, \ldots, \pm 1)$ of the d-**cube** $\mathbf{C}_d$, which is called a square when $d = 2$ and a cube when $d = 3$. Each of the hyperplanes

$$H_i^\pm = \{\mathbf{x} \in \mathbb{R}^d \mid x_i = \pm 1\}, \quad i = 1, 2, \ldots, d,$$

bounds a half-space containing the origin O, and $\mathbf{C}_d$ is the intersection of these $2d$ half-spaces. For $2 \le k \le d-1$, every k-face of a d-cube is a k-cube. $\mathbf{C}_d$ is bounded by its $2d$ faces, and its interior consists of those $\mathbf{x} \in \mathbb{R}^d$ in which every coordinate is less than 1 in modulus. No prizes for guessing the dihedral angles (all equal to $\pi/2$), nor for proving that just d edges meet at any vertex V_0. Any two such edges are transposed by reflection in their perpendicular hyperplanar bisector, which is the hyperplane passing through O, V_0 and the other $d-2$. These transpositions generate a subgroup of $\mathrm{Sym}(\mathbf{C}_d)$

isomorphic to the symmetric group S_d. On the other hand, reflections in the d coordinate hyperplanes generate a copy of Z_2^d, and it can be shown that $\mathrm{Sym}(\mathbf{C}_d) \cong S_d \times_\alpha Z_2^d$, where α is the natural action. These groups comprise one of the four infinite families of finite Coxeter groups.

Our second type of standard regular polytope is obtained from the first as follows. Take a vertex V_0 of $\mathbf{C}_{d+1}$ and consider the $d + 1$ vertices joined to V_0 by an edge. For example, if V_0 is the point with all $d + 1$ coordinates equal to 1, let V_i have ith coordinate equal to -1 and all others equal to 1. Then the points $V_1, \ldots, V_{d+1}$ all lie

- at a distance 2 from V_0,

- at a distance $2\sqrt{2}$ from each other,

- in the hyperplane H given by $\sum_{k=1}^{d+1} x_k = d - 1$ in $\mathbb{R}^{n+1}$.

These $d + 1$ points form the vertex set of a **regular d-simplex $\mathbf{T}_d$**. $\mathbf{T}_2$ is an equilateral triangle and $\mathbf{T}_3$ is a regular tetrahedron. The $d + 1$ faces of $\mathbf{T}_d$ are all copies of $\mathbf{T}_{d-1}$ and every pair of vertices of $\mathbf{T}_d$ are joined by an edge. The dihedral angles of $\mathbf{T}_d$ are harder to calculate than those of $\mathbf{C}_d$. In contrast, it is easy to see that $\mathrm{Sym}(\mathbf{T}_d) \cong S_d$, and these groups form another infinite family of finite Coxeter groups.

The third and final type is the **cocube** (or cross polytope) $\mathbf{C}_d^*$, defined as the dual of $\mathbf{C}_d$. Its vertices are thus the $2d$ points on the coordinate axes at a distance 1 from the origin. $\mathbf{C}_2^*$ is a square (again) and $\mathbf{C}_3^*$ is a regular octahedron. A quick check shows that $\mathbf{C}_d^*$ has 2^d faces, each a copy of $\mathbf{T}_{d-1}$, and that the $2(d - 1)$ vertices adjacent to a given one form the vertex set of a $\mathbf{C}_{d-1}^*$. And, of course, the symmetry group of $\mathbf{C}_d^*$ is the same as that of $\mathbf{C}_d$.

This completes the standard list of regular polytopes, which forms the basis of the classification theorem. We have three regular polytopes in every dimension, and they are genuinely different except in the case $d = 2$, when $\mathbf{C}_2$ and $\mathbf{C}_2^*$ coincide. In addition, we have

(a) when $d = 2$ the regular n-gon for all $n \geq 5$,

(b) when $d = 3$ the exceptional types $\mathbf{D}$ and $\mathbf{I}$,

(c) three further exceptional types when $d = 4$,

(d) nothing else.

All their symmetry groups are Coxeter groups, including the dihedral groups D_{2n} (when $d = 2$) as a third infinite family. Two things remain to be done: to describe the three exceptional types in dimension 4 and to prove that this final list is complete.

12.2 The Exceptional Types in Dimension Four

The construction begins, as usual, with a cube, in this case the 4-cube C_4 with 16 vertices $(\pm 1, \pm 1, \pm 1, \pm 1) \in \mathbb{R}^4$. By doubling the linear dimensions of the inscribed cocube, we obtain a cocube of C_4^* with 8 vertices $(\pm 2, 0, 0, 0)$, $(0, \pm 2, 0, 0)$, $(0, 0, \pm 2, 0)$, $(0, 0, 0, \pm 2) \in \mathbb{R}^4$. These 24 points form the vertex set of our first exceptional type. We denote this polytope by P and its group $\mathrm{Sym}(P)$ by G. For future reference, we record the obvious fact that $\mathrm{Sym}(C_4) \subseteq G$.

The vertex $V = (2, 0, 0, 0)$ has distance 2 from each of the eight vertices $(1, \pm 1, \pm 1, \pm 1)$, which define a 3-cube C. The 1-, 2-, 3-faces of P containing V are thus in one-to-one correspondence with the 8 vertices, 12 edges, 6 faces of C respectively. The 2-faces are triangles and the 3-faces are octahedra.

Moreover, the group $\mathrm{Sym}(C)$ acts regularly on the flags of C and thus on the flags of P containing V, and $\mathrm{Sym}(C) \subseteq G$ as the subgroup of elements fixing V (the **stabiliser** $\mathrm{Stab}(V)$ of V). To prove that P is regular, it remains to check that G acts **transitively** on the vertices of P, that is, for any vertex V' there is a $g \in G$ such that $Vg = V'$. But the subgroup $\mathrm{Sym}(C_4)$ of G acts transitively on the vertices of C_4 and on those of C_4^*, and so it suffices to find an isometry of P that transposes $(2, 0, 0, 0)$ and $(1, -1, -1, -1)$. An obvious choice is reflection in the perpendicular hyperplanar bisector of these two points, which maps $\mathbf{x} = (x, y, z, t)$ to $\mathbf{x} - \mathbf{x}\sigma/2$, where $\mathbf{x}\sigma$ is the vector with all four coordinates equal to $x + y + z + t$:

$$(x, y, z, t) \mapsto \frac{1}{2}(x - y - z - t, -x + y - z - t, -t, -x - y + z - t, -x - y - z + t).$$

Simple counting based on the rules of incidence contained implicitly in the previous paragraph yields the following facts. First, the number of vertices, edges, 2-faces and faces of P is

$$v = 24, \quad e = 96, \quad f = 96, \quad h = 24,$$

respectively. Because of the last equation, the regular polytope P is commonly referred to as the 24-**cell**. Also, these values illustrate Euler's formula

$$v - e + f - h = 0$$

in dimension 4 and suggest (correctly) that P is self-dual. Second, the order of G is obtained as the total number of flags of P, namely,

$$|G| = 24 \cdot 8 \cdot 3 \cdot 2 = 2^7 \cdot 3^2 = 1152.$$

The 24 vertices of the 24-cell P form part of the vertex set of our second exceptional type. The others are all points obtained from $(\pm \tau, \pm 1, \pm \tau^{-1}, 0)$, where $\tau = (1 + \sqrt{5})/2$, by an even permution of the coordinates. Denote this polytope by Q and now let $G = \mathrm{Sym}\, Q$. Then Q has 120 vertices and is shown to be regular by the method used above for P, as follows.

The edges containing the vertex $V = (2, 0, 0, 0)$ are the 12 segments connecting V to the vertices

$$(\tau, \pm 1, \pm \tau^{-1}, 0), \ (\tau, 0, \pm 1, \pm \tau^{-1}), \ (\tau, \pm \tau^{-1}, 0, \pm 1),$$

which form the vertex set of a regular icosahedron I (see Exercise 10.11). $\mathrm{Sym}(I)$ is the stabiliser of V in G and acts transitively on the flags of Q containing V. Routine checking shows that the nine orbits of $\mathrm{Sym}(I)$ on the 120 vertices of Q are the subsets consisting of points with fixed first coordinate 0, $\pm \tau^{-1}$, ± 1, $\pm \tau$, ± 2. Each of the four pairs of consecutive points in the sequence

$$(0, 2, 0, 0), \ (\tau^{-1}, \tau, 1, 0), \ (1, 1, 1, 1), \ (\tau, 1, \tau^{-1}, 0), \ (2, 0, 0, 0)$$

is joined by an edge (of length $2\tau - 2$) of Q. Reflections in the four corresponding perpendicular bisectors, along with reflection in the origin, show that G acts transitively on the vertices of Q. Hence, G acts transitively on the flags of Q, and so Q is indeed regular.

Now for some counting, based on the first sentence of the previous paragraph. The number of vertices, edges, faces of I is 12, 30, 20 respectively, and these are thus the numbers of 1-, 2-, 3-faces of Q containing V. The 2-faces are triangles and the 3-faces are tetrahedra. The total number of k-faces of Q, $k = 0, 1, 2, 3$, is therefore

$$v = 120, \ e = 120 \cdot 12/2 = 720, \ f = 120 \cdot 30/3 = 1,200, \ h = 120 \cdot 20/4 = 600,$$

respectively. These values again illustrate Euler's formula, and the last equation explains the name 600-**cell** for Q. Since each edge of Q is contained in five 2-faces (one for each edge of I containing a given vertex), the total number of flags of Q is given by

$$|G| = 120 \cdot 12 \cdot 5 \cdot 2 = 14,400.$$

The lack of symmetry in the above sequence v, e, f, h for Q shows that Q is not self-dual, and we have our third exceptional type Q^*. Its symmetry group is the same as that of Q, and from the values

$$v = 600, \ e = 1,200, \ f = 720, \ h = 120$$

we get its name, the 120-**cell**.

Details of the six regular polytopes in dimension 4 are summarised in Table 12.1, where the columns headed v, e, f, g are as in Table 10.1 and h is the number of faces. The penultimate column names the polyhedron P' defined by

the vertices joined by an edge to a fixed vertex of the polytope P. More of this
in the next section, where the meaning of the symbol s in the last column will
also be explained.

Table 12.1   The six regular 4-polytopes

P	v	e	f	h	g	P'	s
T_4	5	10	10	5	120	T_3	$\{3,3,3\}$
C_4	16	32	24	8	384	T_3	$\{4,3,3\}$
C_4^*	8	24	32	16	384	C_3^*	$\{3,3,4\}$
24-cell	24	96	96	24	1,152	C_3	$\{3,4,3\}$
120-cell	600	1,200	720	120	14,400	T_3	$\{5,3,3\}$
600-cell	120	720	1,200	600	14,400	I	$\{3,3,5\}$

12.3 Three Concepts and a Theorem

The proof in the next section that our list of regular polytopes is now complete will necessarily require an argument by induction on the dimension. We
therefore need a link between regular polytopes of dimension d and those of
dimension $d - 1$. This is supplied eponymously as follows.

Definition 12.1

For a given regular polytope P, the set of vertices of P joined by an edge to a
fixed vertex defines the **link** P' of P.

While several examples and applications of this concept appear in the previous section, the definition requires some justification. First let O be the point
whose coordinates are the averages of the corresponding coordinates of the vertices of P. Since any $g \in G = \text{Sym}(P)$ permutes these vertices, we must have
$Og = O$. And since this action is transitive, all the vertices of P are equidistant
from O. It therefore makes sense to call O the **centre** of P and the distance
from O to a vertex the **radius** of (the circumsphere of) P. Next let V be any
vertex of P. Then, by transitivity again, all the vertices joined by an edge to V
lie in a hyperplane H perpendicular to the line through O and V and form the
vertex set of the polytope $P' = P \cap H$. The hyperplanes bounding P that contain V intersect H in the hyperplanes that bound P', and so we get one-to-one
correspondences (generically called σ) between the faces of P containing V and

the faces of P', the k-faces of P containing V and the $(k-1)$-faces of P', and the flags of P containing V and the flags of P'. It follows from the regularity of P that $\mathrm{stab}(V) = \{g \in G \mid Vg = V\}$ acts regularly on the flags of P'. Hence, P' is regular, has dimension one less than that of P, and is independent (up to isometry) of the choice of V.

The next definition, which is due to Schläfli, provides the key to the whole business.

Definition 12.2

The **symbol** of a d-dimensional regular polytope P is the sequence $\{r_1, r_2, \ldots, r_{d-1}\}$ of integers defined inductively as follows: r_1 is the number of sides of a 2-face of P and $\{r_2, \ldots, r_{d-1}\}$ is the symbol of the link P' of P.

Starting with $d = 2$, we get the symbol $\{n\}$ for the regular plane n-gon, $n \geq 3$. This gives the symbols

$$\{3,3\}, \quad \{4,3\}, \quad \{3,4\}, \quad \{5,3\}, \quad \{3,5\}$$

for $\mathbf{T} = T_3$, $\mathbf{C} = C_3$, $\mathbf{O} = C_3^*$, $\mathbf{D}$, $\mathbf{I}$ respectively when $d = 3$, and these in turn imply the entries in the last column of Table 12.1 when $d = 4$. Then induction on d yields

$$\{3, \ldots, 3\}, \quad \{4, 3, \ldots, 3\}, \quad \{3, \ldots, 3, 4\}$$

for the symbols of T_d, C_d, C_d^* respectively. These values illustrate the obvious fact that any entry in any Schläfli symbol must be at least 3.

An inductive argument using the correspondence σ referred to above shows that if P has symbol $\{r_1, \ldots, r_{d-1}\}$, then a face of P has symbol $\{r_1, \ldots, r_{d-2}\}$. This leads, again by induction, to three more facts.

Fact 1. *If P has symbol $\{r_1, \ldots, r_{d-1}\}$, then the dual P^* of P has symbol $\{r_{d-1}, \ldots, r_1\}$.* The proof also uses the one-to-one correspondence that exists between the k-faces of P and the $(d-k-1)$-faces of P^*.

Fact 2. *A regular polytope is determined up to equivalence by its symbol.* This means that two regular polytopes with the same symbol differ only in size and position. The proof also uses Theorem 12.1 below.

Fact 3. *The only tessellations of $\mathbb{R}^d$ by regular polytopes are*

(i) *that of $\mathbb{R}^d$ by d-cubes for all d,*

(ii) *the dual pair $(3,6)$, $(6,3)$ of $\mathbb{R}^2$ by hexagons, triangles respectively,*

(iii) *a dual pair in $\mathbb{R}^4$ by cocubes, 24-cells.*

The proof of this uses Theorems 12.1 and 12.2 below.

Definition 12.3

For a regular polytope P with radius r and edge-length l, we set

$$\rho(P) = l^2/4r^2.$$

Theorem 12.1

If P has symbol $\{r_1, \ldots, r_{d-1}\}$, then

$$\rho(P) = 1 - \frac{\cos^2 \pi/r_1}{\rho(P')}.$$

Proof

Let O, r, l denote the centre, radius, edge-length of P respectively. Likewise let O', r', l' denote the centre, radius, edge-length of the link P' of P at a vertex V. Further let V_1, V_2 be the endpoints of an edge of P'; see Fig. 12.1. The points V_1, V, V_2 are consecutive vertices of a 2-face of P, which is a regular r_1-gon in the plane containing these three points. Letting r'' denote the radius of this r_1-gon, we have

$$l = 2r'' \sin \pi/r_1, \qquad l' = 2r'' \sin 2\pi/r_1,$$

whence $l' = 2l \cos \pi/r_1$. Moreover, by definition,

$$\rho(P) = l^2/4r^2, \qquad \rho(P') = l'^2/4r'^2.$$

Letting 2ϕ denote the angle at the apex O of the isosceles triangle V_1OV, we have

$$l = 2r \sin \phi, \qquad r' = l \cos \phi,$$

and we compute

$$\rho(P) = \sin^2 \phi, \qquad \rho(P') = \frac{\cos^2 \pi/r_1}{\cos^2 \phi}.$$

The desired formula now follows by Pythagoras.   □

It follows by induction from this theorem that $\rho(P)$ depends only on the symbol $\{r_1, \ldots, r_{d-1}\}$, and this is a step in the proof of Fact 2 above. The formula can thus be rewritten in the form

$$\rho(r_1, \ldots, r_{d-1}) = 1 - \frac{\cos^2 \pi/r_1}{\rho(r_2, \ldots, r_{d-1})}. \tag{12.1}$$

This leads to the classification of regular tessellations in Fact 3 above and to that of regular polytopes in the next section.

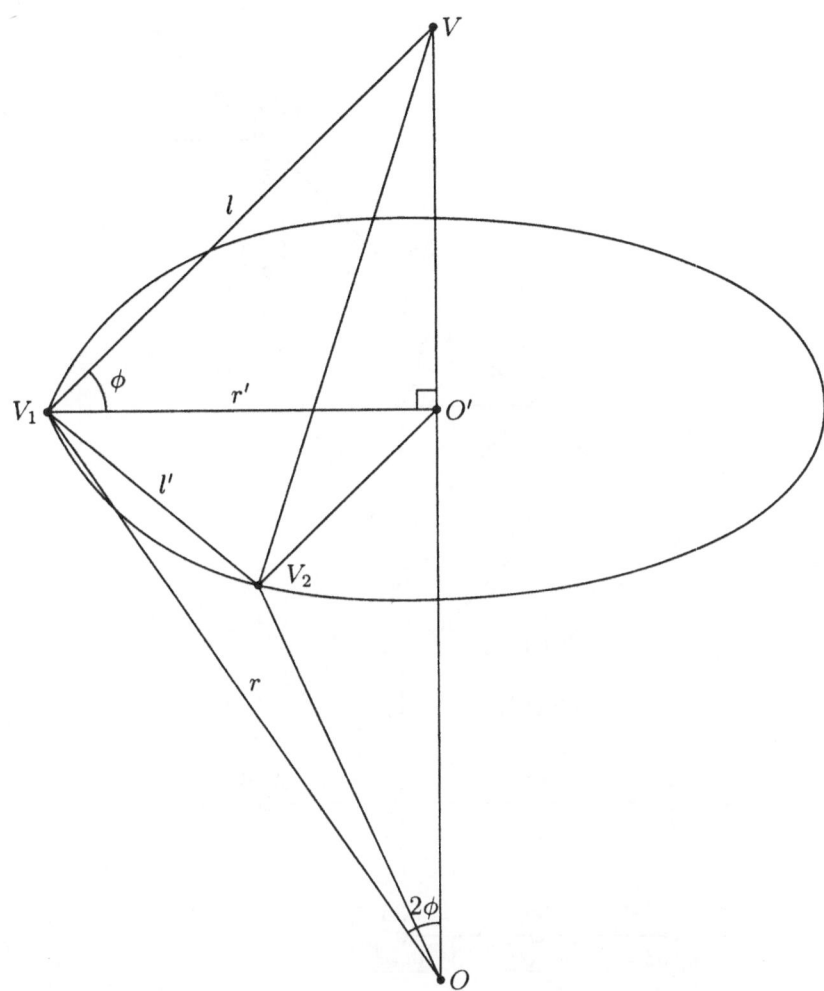

Figure 12.1   A 3-dimensional slice of P

12.4 Schläfli's Theorem

Theorem 12.2

The only possible symbols for a regular polytope of dimension d are

$\{n\}$, *where n is an integer ≥ 3, when $d = 2$,*

$\{3,3\}$, $\{4,3\}$, $\{3,4\}$, $\{5,3\}$, $\{3,5\}$ *when $d = 3$,*

$\{3,3,3\}$, $\{4,3,3\}$, $\{3,3,4\}$, $\{3,4,3\}$, $\{5,3,3\}$, $\{3,3,5\}$ *when $d = 4$,*

$\{3,\ldots,3\}$, $\{4,3,\ldots,3\}$, $\{3,\ldots,3,4\}$ *when $d \geq 5$.*

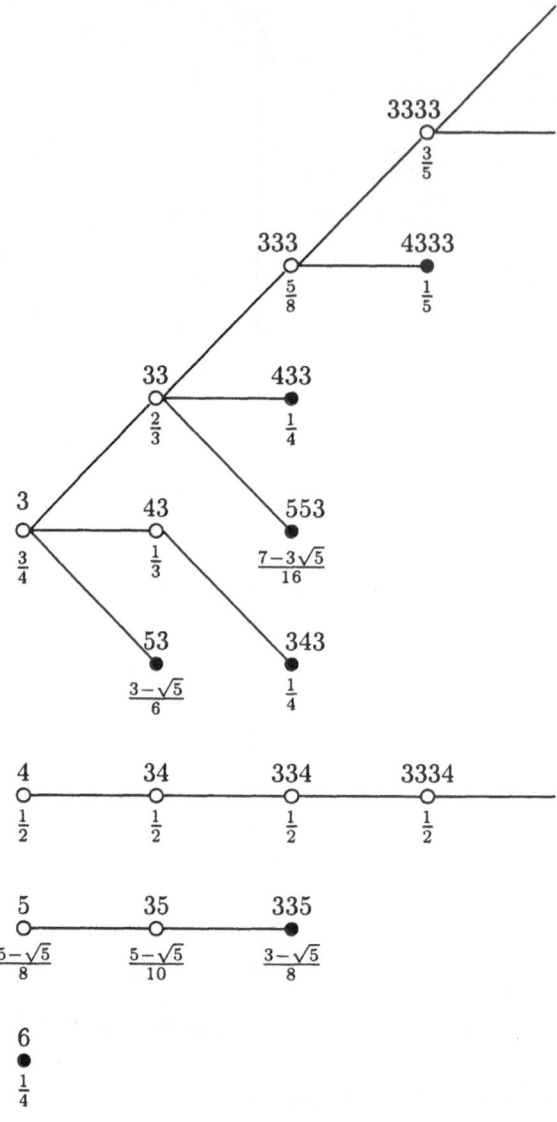

Figure 12.2   Values of ρ

Proof

The idea is to draw up a list of regular polytopes P starting with plane polygons and inductively constructing for each P those polytopes with link P. The entries in the list are represented by the nodes in Fig. 12.2, and the inductive construction by moving along an edge from left to right. The columns correspond to dimensions 2, 3, 4, 5. Above each node is written the symbol (without braces and commas) of the polytope it represents, and below is the corresponding value of ρ (see the Exercises). The situation in $d = 5$ repeats indefinitely for all $d \geq 6$ as indicated, and likewise the first column for $n \geq 7$. The filled-in nodes betoken values of $\rho \leq 1/4$, when further progress is impossible: since $r_1 \geq 3$, the right-hand side of (12.1) is non-positive for such values of $\rho(r_2, \ldots, r_{d-1})$. And there we are. □

EXERCISES

12.1. By choosing coordinates and using an isometry, prove that any hyperplane partitions $\mathbb{R}^d$ into two half-spaces.

12.2. Regarding our three definitions of regularity of a polytope, show that (a) and (b) are consequences of (c).

12.3. Check that definition c) is equivalent to those given earlier in dimensions 2 and 3.

12.4. Prove that every k-face of a regular polytope of dimension $d > k$ is a regular polytope of dimension k.

12.5. Let F_1, F_2 be adjacent faces of a regular polytope P that are defined by hyperplanes H_1, H_2 respectively. Show that reflection in the hyperplane H containing $H_1 \cap H_2$ and the centre of P is a symmetry of P transposing F_1 and F_2.

12.6. What is the common value of the dihedral angles of a 4-cube?

12.7. By counting k-faces and studying their properties of incidence, calculate the number of flags in the 4-cocube.

12.8. Let $\mathbf{T}_d$ be the link of $\mathbf{C}_{d+1}$ at the vertex $(1, 1, \ldots, 1)$ as defined in the text. Find the equations of the hyperplanes in $\mathbb{R}^{d+1}$ whose intersections with the space $H = \{(x_1, x_2, \ldots, x_{d+1}) \mid \sum_{k=1}^{d+1} x_k = d - 1\}$ define $\mathbf{T}_d$.

12.9. Find the coordinates of the vertices of the link of $\mathbf{C}_d^*$.

12.10. Show that the 24-cell is self-dual.

12.11. Check that the given 12 vertices of the link of the 600-cell do indeed define a regular icosahedron. What is the equation of the hyperplane in $\mathbb{R}^4$ that contains it?

12.12. This one is fairly tough. Use duality to find the vertices of the link of the 120-cell in $\mathbb{R}^4$.

12.13. Check the values given for v, e, f, h for each of the exceptional polytopes in dimension 4.

12.14. For each k, $0 \leq k \leq d - 1$, count the number of k-faces of the regular polytopes T_d, C_d, C_d^*, $d \geq 2$. On the basis of this evidence, write down Euler's formula for arbitrary dimension $d \geq 2$.

12.15. Prepare a complete proof that a regular polytope P is uniquely determined by its Schläfli symbol using the following ingredients:

   (i) induction on the dimension,

   (ii) the symbol of a face of P,

   (iii) the fact that $\rho(P)$ depends only on the symbol of P,

   (iv) Exercise 12.4 above.

12.16. It turns out that $\mathbb{R}^d$ can be tiled by copies of $\{r_1, \ldots, r_{d-1}\}$ if and only if there is a regular polytope $\{r_2, \ldots, r_d\}$ such that $\rho(r_2, \ldots, r_d)$ $= \cos^2 \pi / r_1$. Use this hint in conjunction with Theorem 12.2 to classify all regular tessellations of $\mathbb{R}^d$, $d \geq 2$.

12.17. Check the values of ρ appearing in Fig. 12.2, and in the four infinite sequences $(\{n\}, T_d, C_d, C_d^*)$ emanating from it.

Solutions

Chapter 1

1.1 It follows directly from the definition of the modulus function on $\mathbb{R}$ (see Example 1.2) that $|x| \geq 0$ for all x, and $|x| = 0 \Leftrightarrow x = 0$. Axiom M1 follows. For M2, it suffices to check that $|-x| = |x|$ for all x. Finally, M3 follows from the inequality

$$|a + b| \leq |a| + |b|,$$

which is an equality if (i) either a or b is zero, or (ii) a and b have the same sign. If (iii) a and b have opposite signs, then strict inequality holds, since the sum of two positive numbers exceeds their difference.

1.4 Take the discrete metric on the set $\mathbb{Z}$ and consider the map

$$\mathbb{Z} \to \mathbb{Z}, \quad n \mapsto 2n.$$

1.6 This is a consequence of the following more general assertion: if A, B are sets with $|A| = |B| = n \in \mathbb{N}$, then the number s_n of bijections from A to B is $n!$. This is proved by induction on n, with base $s_1 = 1 = 1!$. Let $n \geq 2$, assume the result for $n - 1$, and fix an element $a \in A$. Then for each of the n elements $b \in B$, there are s_{n-1} bijections $\beta: A \to B$ sending a to b, by the inductive hypothesis applied to the sets $A \setminus \{a\}$, $B \setminus \{b\}$. It follows that $s_n = n \cdot (n-1)! = n!$, as required.

1.7 Both metrics are preserved by an arbitrary translation,

$$t(\mathbf{a}): \mathbb{R}^2 \to \mathbb{R}^2, \quad \mathbf{x} \mapsto \mathbf{x} + \mathbf{a},$$

where $\mathbf{a} \in \mathbb{R}^2$. The Euclidean metric is preserved by any reflection $r(l)$, where l is a line in $\mathbb{R}^2$. But $r(l)$ preserves the Manhattan metric only if l is parallel to one of the coordinate axes or one of the lines $y = \pm x$. For example, if l is the line $y = x/\sqrt{3}$, the image of the point $(1, 0)$ under $r(l)$ has coordinates $(1/2, \sqrt{3}/2)$, and $1 + 0 \neq 1/2 + \sqrt{3}/2$.

1.9 Suppose that
$$u = r^\varepsilon t(c) = r^\delta t(d),$$
where $\varepsilon, \delta \in \{0, 1\}$ and $c, d \in \mathbb{R}$. Then $0u = c = d$. Thus, $c = d$ and cancellation of $t(c) = t(d)$ gives $r^\varepsilon = r^\delta$, whence $\varepsilon = \delta$ and the form is unique.

1.11 Let $a, b \in \mathbb{R}$, $a \neq b$, and let u be an isometry of $\mathbb{R}$ such that $au = a$, $bu = b$. Then x, xu are equidistant from a and from b. Since there are at most two points of $\mathbb{R}$ equidistant from a given point, the assumption that $xu \neq x$ leads to the contradiction that $xu = 2a - x = 2b - x$. So $xu = x$ and $u = 1$. For the rest, if u, v are isometries having the same effect on a and b, then uv^{-1} fixes a and b. Then $uv^{-1} = 1$ and $u = v$.

1.12 The only possibilities are 1, 2, ∞.

1.14 $(x^y)^z = z^{-1}(y^{-1}xy)z = (z^{-1}y^{-1})x(yz) = x^{(yz)}$ as $(yz)^{-1} = z^{-1}y^{-1}$. The relation is reflexive as $g^1 = g$ for all x in the group. For symmetry,
$$g^x = h \Rightarrow h^{x^{-1}} = (g^x)^{x^{-1}} = g^{xx^{-1}} = g^1 = g.$$
Finally, if $g^x = h$ and $h^y = k$, then $g^{xy} = h^y = k$, and the relation is transitive.

1.16 Setting $u = r^\varepsilon t(c)$, $v = r^\delta t(d)$ as in (1.9), we compute
$$\begin{aligned}
[u, v] &= (vu)^{-1}(uv) = (r^{\delta+\varepsilon}t(c + (-1)^\varepsilon d))^{-1}r^{\varepsilon+\delta}t(d + (-1)^\delta c) \\
&= t(-c - (-1)^\varepsilon d)r^{-\delta-\varepsilon}r^{\varepsilon+\delta}t(d + (-1)^\delta c) \\
&= t(d(1 \pm 1) - c(1 \pm 1)).
\end{aligned}$$

Thus, every commutator is a translation and (taking $c = 0$, $\varepsilon = 1$) vice versa. Thus $G' = T = G^+$.

1.18 Obviously because it is false, but why? Checking the effects of $u^{-1}t(c)u$ and $t(cu)$ on an arbitrary point xu in $\mathbb{R}$, we get
$$\begin{aligned}
xu &\mapsto xt(c)u = (x + c)u \quad \text{and} \\
xu &\mapsto xu + cu \neq (x + c)u
\end{aligned}$$
in general, since u is not linear.

1.19 Sym($\mathbb{Q}$) consists of isometries $r^\varepsilon t(c)$ with $\varepsilon \in \{0,1\}$ and $c \in \mathbb{Q}$, and is not discrete (consider the points $Ot(1/n)$, $n \in \mathbb{N}$). Sym(± 1) $= \{1, r\}$ and is discrete.

1.20 The presentation is $\langle x, y \mid x^2 = y^2 = 1 \rangle$.

Chapter 2

2.2 Note first that if M is the midpoint of the segment AB and C is any point of l, then triangles AMC, BMC are congruent (SAS) so that $d(A, C) = d(B, C)$ for all $C \in l$. For the converse, let M be the origin of coordinates and the line through A, B the x-axis. Then the x-coordinate of any point P on the same side of l as A has the same sign as that of A. Hence $d(A, C) < d(B, C)$, and the reverse inequality holds for points on the same side of l as B. So $d(A, P) = d(B, P) \Rightarrow P \in l$. For the rest, let l be any line in $\mathbb{R}^2$ and A any point of $\mathbb{R}^2 \setminus l$. Then l is the perpendicular bisector of A and its image $B = Ar(l)$ under reflection in l. Then for any isometry u of $\mathbb{R}^2$, it is clear that lu is the perpendicular bisector of Au and Bu, and so it is a line.

2.3 m denotes magnification, say $(x, y)m = (2x, 2y)$, sending lines to parallel lines, circles to circles, and triangles to similar triangles. Angles are preserved but distances are doubled.

2.6 It turns out that the only linear isometries are those that fix the origin O, that is, rotations about O and reflection in lines through O.

2.7 If $P = (x, y)$ has polar coordinates (ρ, θ), then $Ps(\phi)$ has polar coordinates $(\rho, \theta + \phi)$, where $\theta + \phi$ is calculated mod 2π.

2.8 The answer is 6, 2, 1 respectively.

2.9 Choose a point $O \in l$ as the origin of coordinates in which l is the x-axis and put $A = (1, 0)$, $B = (0, 1)$. Then by Exercise 2.4 above (whose solution is contained in the proof of Theorem 2.2), any OP symmetry u of l is determined by Ou, Au. The restriction Sym$^+(l) \to$ Isom($\mathbb{R}$), $u \mapsto u'$, is thus an injection, clearly a homomorphism, and a surjection (consider translations b with axis l and rotations $s = s(P, \pi)$ with $P \in l$). The elements of Sym$^-(l)$ are then of the form $r(l)u$, namely, glide reflections $r(l)t$, and reflections $r(l)s$ in lines perpendicular to l.

2.10 By Theorem 2.2, any $u \in \mathbb{E}$ is determined by its effect on three non-collinear points, A, B, C say in anticlockwise order. In the image under u of the triangle ABC, the vertices Au, Bu, Cu occur anti-clockwise or clockwise but not both. Hence $\mathbb{E}^+ \cup \mathbb{E}^- = \mathbb{E}$ and $\mathbb{E}^+ \cap \mathbb{E}^- = \emptyset$. Under post-multiplication by r we have $\mathbb{E}^+ r \subseteq \mathbb{E}^-$, $\mathbb{E}^- r \subseteq \mathbb{E}^+$. Post-multiplication by r again gives the reverse inclusions in the reverse order. Thus $\mathbb{E}^+ r = \mathbb{E}^-$. The proof that $r\mathbb{E}^+ = \mathbb{E}^-$ is entirely analogous.

2.11 $u = rs(0, \pi)t(2, -1)$, where r denotes reflection in the x-axis.

2.12 The elements $u \in \mathbb{E}$ that lie in $\mathrm{Sym}(O)$ are those with normal form $u = r^\varepsilon s$ with respect to O and l, where l is a line through O, $s \in S_0$ and $r = r(l)$. If $u = r^\varepsilon s(O, \theta)$ and $v = r^\delta s(O, \phi)$, then $uv = r^{\varepsilon+\delta} s(O, \phi + (-1)^\delta \theta)$.

2.13 Take O as the origin of coordinates and the axis of t through O as the x-axis oriented so that the x-coordinate of Ot is positive. If $s = s(O, \phi)$, then the point P with polar coordinates (ρ, θ) given by $\theta = \pi/2 - \phi/2$, $\rho = d(O, Ot)/2 \sin \phi/2$ is mapped to $(\rho, \pi/2 + \phi/2)$ by s, and this point returns to P under t. So $Pst = P$. For the rest, the normal form theorem (with respect to O and any l with $O \in l$) implies that every element of $\mathbb{E}^+$ has the above form st. If $s = 1$ or $t = 1$, this is a translation or a rotation. If not, the first part gives a point P fixed by st. Then the normal form theorem (with respect to P and any l with $P \in l$) shows that st is a rotation about P.

2.14 Apply Theorem 2.4 with $\delta = \varepsilon = 1$, $\alpha = \beta$, $\mathbf{a} = \mathbf{b}$ to $u = r^\delta s(\alpha)t(\mathbf{a})$ to get $u^2 = r^\eta s(\gamma)t(\mathbf{c})$, where $\eta = 0$, $\gamma = 0$, $\mathbf{c} = \mathbf{a}rs(\alpha) + \mathbf{a}$, whence $u^2 = t(\mathbf{c}) \in \mathbb{T}$.

Chapter 3

3.1 Induction on n: the result is trivial for $n = 1, 2$ and guaranteed by the (ordinary) associative law for $n = 3$. So let $n \geq 4$ and assume the result for all k, $1 \leq k < n$. Let p_1, p_2 be two products of $x_1, x_2, \ldots, x_n$ in this order, where the last multiplication to be carried out is that between the first i, j terms and the rest respectively. Assuming without loss of generality that $i < j$, the products $a = x_1 \cdots x_i$, $b = x_{i+1} \cdots x_j$, $c = x_{j+1} \cdots x_n$ are independent of the brackets by hypothesis, and the same goes for the products ab, bc. Then, by the (ordinary) associative law, $p_1 = a(bc) = (ab)c = p_2$, as required.

3.4 The element 2 has no inverse: $2x = 1 \Rightarrow x = 1/2 \notin \mathbb{N}$.

3.5 The function $f: \mathbb{R} \to \mathbb{R}^+$ given by $f(x) = e^x$ is

   (i) injective, since $f'(x) > 0$, $\forall x \in \mathbb{R}$,

(ii) surjective since f is continuous, $\lim\limits_{x \to \infty} f(x) = \infty$ and $\lim\limits_{x \to -\infty} = 0$, and

(iii) a homomorphism since $f(x+y) = f(x)f(y)$.

Thus $f: \mathbb{R} \to \mathbb{R}^+$ is the required isomorphism.

3.8 Since $(xy)\phi = (x\phi)(y\phi) \ \forall x, y \in G$, we have $1\phi = 1$ (take $x = y = 1$) and $(x^{-1})\phi = (x\phi)^{-1}$ (take $y = x^{-1}$). It now follows easily that Ker ϕ, Im ϕ are closed under multiplication and inversion and contain the identity element, and thus are subgroups of G, H respectively. Moreover, for all $k \in K$, $g \in G$ we have $(g^{-1}kg)\phi = (g^{-1})\phi(k\phi)(g\phi) = (g\phi)^{-1}1(g\phi) = 1$, so that $g^{-1}kg \in K$ and $K \lhd G$. It is a matter of routine to check that ϕ' is well defined (that is, independent of the choice of coset representative x), injective, surjective and multiplicative, and is thus an isomorphism.

3.9 For $x \in G$, $x \in$ Ker $\gamma \Leftrightarrow \gamma_x = 1 \in \mathrm{Aut}(G) \Leftrightarrow \forall g \in G \ g = g\gamma_x = x^{-1}gx \Leftrightarrow x \in Z(G)$. For any $x \in G$, $\alpha \in \mathrm{Aut}(G)$, $g \in G$ we have

$$g\alpha^{-1}\gamma_x\alpha = (x^{-1}(g\alpha^{-1})x)\alpha = (x\alpha)^{-1}g(x\alpha) = g\gamma_{x\alpha},$$

so that $\alpha^{-1}\gamma_x\alpha = \gamma_{x\alpha} \in \mathrm{Inn}(G)$, whence $\mathrm{Inn}(G) \lhd \mathrm{Aut}(G)$.

3.12 Since $w(X)$ contains 1 (the empty product) and is closed under multiplication and inversion, it is a subgroup of G. Since $w(X)$ contains X (products of length 1), it is one of the terms in the intersection defining $\langle X \rangle$. Hence $\langle X \rangle \subseteq w(X)$. On the other hand, $\langle X \rangle$ is a subgroup of G containing X, and so contains all finite products of elements of X and their inverses, that is, $\langle X \rangle \supseteq w(X)$. Thus $w(X) = \langle X \rangle$.

3.13 Since H is a subgroup and $n \in H$, it is clear that $n\mathbb{Z} \subseteq H$. For a typical element $h \in H$, we have $h = 0 \in n\mathbb{Z}$ or $h = \pm m$, $m \in \mathbb{N}$. Then $m \in H$ and we have $m = qn + r$, where q, r are non-zero integers and $0 \leq r < n$, by Euclid's theorem. It follows that $r = m - qn \in H$, and then $r = 0$ by minimality of n. So $h = \pm m = \pm qn \in n\mathbb{Z}$, and $H \subseteq n\mathbb{Z}$, as required.

3.14 Let H be an arbitrary subgroup of Z_n. If H is the trivial group, take $d = n$. Otherwise, let d be the smallest positive integer for which $x^d \in H$. Then, as in the solution to Exercise 3.13, $\langle x^d \rangle \subseteq H$ and d is a divisor of n. A second application of Euclid's theorem shows that every $h \in H$ is a power of x^d. This gives the reverse inclusion, so $H = \langle x^d \rangle$ as required.

3.15 It is a matter of routine to show that $\sim$ is an equivalence relation and to describe the classes. To get left cosets, take the relation $x \sim y$ if and only if $x^{-1}y \in H$.

3.16 For the first part, it is required to show that $\rho_x : \widehat{G} \to \widehat{G}$ is a bijection for any $x \in G$. For any $g \in \widehat{G}$ we have

$$g(\rho_x \rho_{x^{-1}}) = (g\rho_x)\rho_{x^{-1}} = (gx)x^{-1} = g(xx^{-1}) = g1,$$

whence $\rho_x \rho_{x^{-1}} = 1$. Then $\rho_{x^{-1}} \rho_x = 1$ and $\rho_{x^{-1}} = (\rho_x)^{-1}$. So ρ is a bijection. Next let $x, y \in G$ and $g \in \widehat{G}$. Then

$$g(\rho_x \rho_y) = (g\rho_x)\rho_y = (gx)y = g(xy) = g\rho_{xy},$$

whence $\rho_x \rho_y = \rho_{xy}$ for all $x, y \in G$ and ρ is a homomorphism. Finally, if $\rho_x = \rho_y$ for $x, y \in G$, then

$$x = 1x = 1\rho_x = 1\rho_y = 1y = y,$$

which proves that ρ is injective.

3.17 It is obvious that $a1 = a$, $ax = a = ay \Rightarrow axy = a$, $ax = a \Rightarrow ax^{-1} = a$, so $\mathrm{Stab}_G(a) \le G$. The map $\alpha : G \to aG$, $x \mapsto ax$, is surjective by definition, and for $x, y \in G$,

$$x\alpha = y\alpha \Leftrightarrow ax = ay \Leftrightarrow axy^{-1} = a \Leftrightarrow xy^{-1} \in \mathrm{Stab}_G(a).$$

Referring to Exercise 3.15, this means that two elements of G have the same effect on a if and only if they belong to the same right coset of $\mathrm{Stab}_G(a)$ in G. Hence, $|aG| = |G : \mathrm{Stab}_G(a)|$.

3.19 The map $Hx \mapsto x^{-1}H$ is a bijection.

3.20 If T_1, T_2 are right transversals for H in K, K in G respectively, then $T = \{t_1 t_2 \mid t_1 \in T_1, t_2 \in T_2\}$ is a right transversal for H in G.

3.23 Calculations show that both D'_∞ and D'_{2n} are generated by $[y, x] = y^{-2}$: $D'_\infty = \langle y^2 \rangle = D'_{2n}$. So $D^{ab}_\infty \cong Z_2 \times Z_2 \cong D^{ab}_{2n}$ for n even, and $D^{ab}_{2n} \cong Z_2$ for n odd.

3.24 Let x be a suitable reflection and y a rotation through minimal angle, then take $H = \langle x \rangle$.

3.26 Assume first that $h^k = h$ for all $h \in H$, $k \in K$. Let $k \in K$ and $g = k'h \in G$ be arbitrary. Then $k^g = (k^{k'})^h = k^{k'} \in K$, whence $K \lhd G$. Next assume that $K \lhd G$ and let $h \in H$, $k \in K$ be arbitrary. Then, bracketing the commutator $[h, k]$ in two different ways, $(k^{-1})^h k \in K$ and $h^{-1}h^k \in H$. Thus $[h, k] \in H \cap K = \{1\}$, $h^{-1}k^{-1}hk = 1$, and $h^k = h$, as required. It then follows that the map $G \to K \times H$, $kh \mapsto (k, h)$, is an isomorphism.

3.28 For any $P \in \mathbb{R}^2$, PD is a lattice whose cells are unit squares. Then the circle centre P radius $1/2$ contains no point of PD other than P.

3.30 If $Z_2 = \langle x \mid x^2 = 1 \rangle$ and $Z_3 = \langle y \mid y^3 = 1 \rangle$, then the first six powers of $(x, y) \in Z_2 \times Z_3$ are all distinct. If $Z_6 = \langle z \mid z^6 = 1 \rangle$, it follows that the map $Z_2 \times Z_3 \to Z_6$, $(x, y) \mapsto z$ is an isomorphism.

3.31 If $Z_8 = \langle x \mid x^8 = 1 \rangle$, its four automorphisms are induced by mapping $x \to x$, x^3, x^5, x^7 respectively. The pairwise non-isomorphism is proved by counting the elements of order 2 in each of the four resulting groups.

Chapter 4

4.2 (a) The centre of s lies on the axis of r.

(b) The axis of r is perpendicular to the axis of t.

4.3 Given one such decomposition $q = rt$, we know that r and t commute. Hence, $q^2 = (rt)^2 = r^2 t^2 = t^2$ is a translation (cf. Exercise 2.14). So if $q^2 = t(\mathbf{a})$, then $t = t(\frac{1}{2}\mathbf{a})$. This means that t is determined uniquely by q, and then so is $r = qt^{-1}$.

4.5 Let l and m be the lines through P, P' and Q, Q' respectively and write $p = r(l)t$, $q = r(m)t'$, where $t, t' \in \mathbb{T}$. Since $r(l)$ and t commute, we have $pq = r(l)tr(m)t' = tr(l)r(m)t'$. Then $pq \in \mathbb{T}$ if and only if $t^{-1}(pq)t'^{-1} \in \mathbb{T}$, that is, $r(l)r(m) \in \mathbb{T}$. By Theorem 4.1, this holds if and only if $l\|m$.

4.7 Let s be a rotation of $\mathbb{R}^3$ with axis l. If H is any plane perpendicular to l, then s acts on H as a rotation $s(H)$ with centre $O = l \cap H$. By the converse of Theorem 4.1, there are lines $l_1, l_2 \in H$ such that $s(H) = r(l_1)r(l_2)$, and then $s = r(H_1)r(H_2)$, where H_1 and H_2 are the planes containing l, l_1 and l, l_2 respectively.

For the rest, let s, s' be rotations of $\mathbb{R}^3$ with axes l, l' respectively. If $l = l'$ then ss' is again a rotation about this axis, so assume from now on that $l \neq l'$. Suppose first that l and l' intersect, in a point O say, and let H, H' be the planes through O perpendicular to l, l' respectively. Then, as in the first part of the exercise, there are planes H_1, H_2 and H_1', H_2' through l and l' respectively such that $s = r(H_1)r(H_2)$ and $s' = r(H_1')r(H_2')$. By latitude of choice, we can take both H_2 and H_1' to be the plane H' containing l and l', whereupon $ss' = r(H_1)r(H_2')$ is a rotation with axis $H_1 \cap H_2'$. (Note that $H_1 \cap H_2'$ really is a line, since it contains O and $H_1 \neq H_2'$ unless $ss' = 1$.) When l and l' are parallel and coplanar, the same argument works, except that now it may happen that $H_1 \cap H_2' = \emptyset$, whereupon ss' is a translation of $\mathbb{R}^3$.

4.8 Let $u \in \mathbb{E}$, with $u \neq 1$ to avoid triviality. Let A be any point of $\mathbb{R}^2$ with

$Au \neq A$, l the perpendicular bisector of A and Au, and $r_1 = r(l_1)$. Then ur_1 fixes A. If $ur_1 = 1$, then $u = r_1$. Otherwise, let $B \in \mathbb{R}^2$ be such that $Bur_1 \neq B$, l_2 the perpendicular bisector of B and Bur_1, and $r_2 = r(l_2)$. Then ur_1r_2 fixes B and, since $A \in l_2$, also A. If $ur_1r_2 = 1$, stop. Otherwise, let C be a point such that $Cur_1r_2 \neq C$, l_3 the perpendicular bisector of C and Cr_2r_1u, and $r_3 = r(l_3)$. Then $ur_1r_2r_3$ fixes three non-collinear points, and so is equal to 1. The result follows.

4.10 Use the normal form theorem (with respect to O and any $l \ni O$) and write s (if it appears) as the product of two suitably chosen reflections.

Chapter 5

5.1 This is analogous to Exercise 3.12. Since the set S^G is closed under conjugation by elements of G, so is the subgroup $\overline{S}$ it generates, that is, $\overline{S}$ is normal in G. Since $S \subseteq S^G \subseteq \overline{S}$, $\overline{S}$ is one of the terms in the intersection I, and so $I \subseteq \overline{S}$. On the other hand, $S \subseteq I$ by definition, so $S^G \subseteq I$ as $I \lhd G$, and then $\overline{S} \subseteq I$ as I is a subgroup. Hence $I = \overline{S}$.

5.2 By Euclid's theorem, there are $s, t \in \mathbb{Z}$ such that $sk + tl = h$.

5.3 Put $r = x^{-1}yxy$, $s = y^{-1}xyx$ and read anti-clockwise around the boundary of the diagram from the NE corner:

$$\begin{aligned}
y^4 &= (yx^{-1}y^{-1}x^{-1})xy(xyxy^{-1})(yx^{-1}yx)(x^{-1}yxy)y^{-1}x^{-1} \\
&= (s^{-1})^{y^{-1}}xy(s^{y^{-1}})(r^{y^{-1}})(r)y^{-1}x^{-1} \\
&= (s^{-1})^{y^{-1}}s^{y^{-2}x^{-1}}r^{y^{-2}x^{-1}}r^{y^{-1}x^{-1}}.
\end{aligned}$$

5.5 From the last relation, $y^{x^2} = (y^x)^x = (y^2)^x = (y^x)^2 = (y^2)^2 = y^4$, and similarly $y^{x^3} = y^8$. But $y^{x^3} = y^1 = y$ by the first relation. Hence, $y^7 = 1$ and we deduce that $y = 1$ using the second relation. So $G = \langle x \mid x^3 = 1 \rangle$ by Tietze transformations, whence $|G| = 3$ and not 15.

5.6 The first part, showing that $\sim$ is an equivalence relation, is routine. For any $i \in \Omega_n$, consider the sequence of points i, $i\sigma, \ldots, i\sigma^k, \ldots$ in the σ-orbit of i. Since Ω_n is a finite set, these points cannot all be distinct: let $i\sigma^l$ be the first duplicate, so that the points $i\sigma^j$, $0 \leq j \leq l-1$, are all distinct and $i\sigma^l$ is equal to one of them. Since σ is one-to-one, we must have $i\sigma^l = i$, and then the points i, $ir, \ldots, i\sigma^{l-1}$ constitute the orbit O_i of σ on i and σ acts on them as a cycle $\gamma_i = (i\ i\sigma \cdots i\sigma^{l-1})$. Note that $\gamma_i = \gamma_j$ for all $j \in O_i$. Now define integers $m(i)$, $i \geq 1$, inductively as follows: $m(1) = 1$ and, for $k > 1$, $m(k)$ is the smallest positive integer not in the set $\bigcup\limits_{j=1}^{k-1} O_{m(j)}$. Since $1 = m(1) < m(2) < \cdots < n$, the sequence of

the $m(k)$ is finite, and their definition guarantees that the orbits $O_{m(k)}$ are pairwise disjoint. Then $\sigma = \gamma_{m(1)}\gamma_{m(2)} \cdots$ is the required decomposition. This decomposition is unique up to the order of the factors.

5.7 $(a_1 \ a_2 \cdots a_l) = (a_1 \ a_2)(a_1 \ a_3) \cdots (a_1 \ a_l)$.

     $(ij) = (1i)(1j)(1i)$.

     $(1j) = (12)(23) \cdots (j-2 \ j-1)(j-1 \ j)(j-2 \ j-1) \cdots (23)(12)$.

5.10   We get a presentation of S_4 on generators a, b, x, z defined by the relations $a^2 = b^2 = (ab)^2 = 1$, $x^3 = 1$, $a^x = b$, $b^x = ab$, $z^2 = 1$, $a^z = a$, $b^z = ab$, $x^z = x^{-1}$. The first three define $Z_2 \times Z_2$ on a, b, the fourth, seventh and tenth define S_3 on x, z, and the other four define the action of the latter on the former.

5.13   The relations $a^{-l} = b^{-m} = (ba)^{-n} = 1$ are equivalent to those in (5.20). Take $Z_2 = \langle y \mid y^2 = 1 \rangle$ with conjugation by y given by the action of α. Then put $x = ay$, $z = yb$ and show that the relations defining the resulting semidirect product are equivalent to those in (5.20).

5.16   If G is abelian, $\tau(G)$ is a subgroup because $1 \in \tau(G)$ as $1^1 = 1$, $x^n = 1 \Rightarrow$ $(x^{-1})^n = 1$, $x^n = 1 = y^m \Rightarrow (xy)^{mn} = x^{nm}y^{mn} = 1$. If $G = D_\infty = \langle r, t \mid$ $r^2 = 1, r^{-1}tr = t^{-1} \rangle$ (see (1.16)), $r, rt \in \tau(G)$ but $rrt = t \notin \tau(G)$.

5.18   Given a matrix with rows y_i, consider the row operations $P(i,j)$ of transposing y_i and y_j, $M(i)$ of multiplying y_i by -1, and $A(i,j)$ of replacing y_i by $y_i + y_j$, where all other rows are left fixed in each case. Then one readily checks that

$$P(i,j) = A(i,j)M(i)A(j,i)M(j)A(i,j)M(i).$$

Allowing **A**-operations to include $y_i \mapsto y_i + ly_k$, $l \in \mathbb{Z}$, the number of factors can be reduced to four.

5.19   In the group $G = \langle x \mid x^m = 1 \rangle \times \langle y \mid y^n = 1 \rangle$, it is easy to see that the element (x, y) has order $l = mn/(m, n)$, the l.c.m. of m and n, and every element has order dividing l. Thus, l is the maximal order of a cyclic subgroup of G. Hence, G is cyclic if and only if $l = mn$, that is, $(m, n) = 1$.

5.20   This follows from the Basis Theorem using the fact that the row operations used to reduce E to Smith normal form preserve the determinant up to sign.

Chapter 6

6.1 Both $g^{-1}sg$ and $s(Og, \pm\theta)$ are OP and fix the point Og, and so both are rotations about this point. Put $g = r^{\varepsilon}s't$ in normal form with respect to O and any $l \ni O$, so that s' fixes O and commutes with s. When $\varepsilon = 0$ we must show that $t^{-1}st = s(Ot, \theta)$, and this is a consequence of the equality of the angles $P\widehat{O}Ps$ and $Pt\widehat{O}tPst$. The case $\varepsilon = 1$ follows from this using the fact that $s^r = s(O, -\theta) = s^{-1}$.

6.3 Let G be any finite subgroup of $\mathbb{E}$. Since $\mathbb{T}$ has no element of finite order other than 1, $G \cap \mathbb{T} = \{1\}$ and so G fixes a point by Theorem 6.1. For the converse, suppose that $G \leq \mathbb{E}$ is discrete and fixes a point O. Then for any point $P \neq O$, its images Pg, $g \in G$, are all equidistant $d(O, P)$ from O and thus lie in the circle centre P radius $3d(O, P)$. Hence the set PG is finite. But if $Pg_1 = Pg_2$, then $g_1 g_2^{-1}$ fixes P as well as O and is thus equal to 1 or $r(l)$, where l is the line through O and P. Then $|G| \leq 2|PG|$ is finite.

6.4 For the purposes of the definition, "circles" in $\mathbb{R}^1$ are just open intervals. If $\mathbb{T}$ is the translation subgroup of $\mathrm{Isom}(\mathbb{R}^1)$ and $G \leq \mathrm{Isom}(\mathbb{R}^1)$ is discrete, first check that $G \cap \mathbb{T} \cong \mathbb{Z}$ or $\{1\}$. Next observe that $G^+ = G \cap \mathbb{T}$ has index 1 or 2 in G. Finally conclude that $G \cong \{1\}$, Z_2, Z or D_∞.

6.5 It is torsionfree of rank at most n.

6.6 Consider a triangle $\Delta = OAB$ with $d(O, A) \geq d(O, B)$ and let C be the circle centre O through A. If P is any point in or on Δ, let the line OP produced meet the edge AB in Q and the circle C in R. Then $d(O, P) \leq d(O, Q) \leq d(O, R) = d(O, A)$. The bound is achieved when $P = A$ (or $P = B$ when $d(O, A) = d(O, B)$).

6.7 For D_{2n}, let C consist of the vertices of a regular plane n-gon. For Z_n, equip C with an orientation, say by adjoining n judiciously chosen points.

6.8 If $G = G^+$, there is nothing to do. If not, let $r \in G \setminus G^+$. Then for any $g \in G$ either g or gr belongs to G^+. Hence $G = G^+ \cup G^+ r^{-1}$ and $|G : G^+| = 2$.

6.9 Suppose that $G \leq \mathbb{E}$ and $T = G \cap \mathbb{T} \cong Z$. Then $T = \langle a \rangle$ for some a and is obviously discrete. For the same reason (counting images of a point in a circle) it is sufficient to prove that $|G : T|$ is finite. By the previous exercise, we need only show that $|G^+ : T|$ is finite. If $G^+ \neq T$ let $g \in G^+ \setminus T$. Then g is a rotation, centre O say. By the corollary to Theorem 2.4, we must have $g = s(O, \pi)$. Thus for any $g' \in G^+ \setminus T$ it follows by preservation of angle (see Fig. 4.5 in Chapter 4) that $g'g^{-1} \in T$, whence $G^+ = T \cup Tg$. So $|G^+ : T| \leq 2$, as required.

6.13 The four types are Z, $Z \times Z_2$, D_∞, $D_\infty \times Z_2$.

6.15 Suppose the squares in Fig. 6.4 have side 1, let O be the centre of one of them, and take the line l through the centres as the x-axis. Then obvious members of $G = \mathrm{Sym}(L)$ are $t = t(1,0)$, $s = (0, \pi)$, $r = r(l)$. Bearing in mind Exercise 6.9, we have $G \cap \mathbb{T} = \langle t \rangle$ and $|G : G^+| = |G^+ : G \cap \mathbb{T}| = 2$. Then it is clear that $G \cong F_2^1$.

Chapter 7

7.1 Proceed as in the solution of Exercise 6.9 and use the fact that the conjugate of a translation by a rotation is a translation through the same distance.

7.3 For each n they are $s^k a^i b^j$, where $i, j, k \in Z$ and $0 \leq k < n$. Such an element has finite order if and only if $k \neq 0$.

7.5 It is sufficient to compare either possible finite orders for elements or indices of the derived groups.

7.6 Since a non-trivial translation and a non-trivial rotation cannot commute, $Z(G_n) = 1$ unless $n = 1$. $Z(G_1) = G_1$ as G_1 is abelian.

7.7 As G_1 is abelian, $G' = \{1\}$ and $G/G' \cong G \cong Z^2$. For $n = 2$, 3, 4, 6, the relation matrices for the presentations in Theorem 7.1 show that the derived factor groups are Z_2^3, Z_3^2, $Z_2 \times Z_4$, $Z_2 \times Z_6$ respectively, and $G'_n \cong Z^2$ in each of these cases.

7.8 G_2 is the exception by the result of the previous exercise. For the others, $G_1 = \langle a, b \rangle$ and $G_n = \langle a, s \rangle$ when $n = 3$, 4, 6. In the last three cases, the 2-generator presentation is given by substituting out the spare generator $b = aa^s$, a^s, a^s respectively using Tietze transformations.

Chapter 8

8.2 From the elementary theory of abelian groups in Chapter 3, and a bit of linear algebra, the following statements are all equivalent:

$G = \langle a', b' \rangle$,

the homomorphism $\phi: G \to G$ induced by $a \mapsto a'$, $b \mapsto b'$ is surjective,

ϕ is bijective,

the matrix $A = \begin{pmatrix} i & j \\ k & l \end{pmatrix}$ has an inverse,

A is unimodular, that is, $\det A = \pm 1$.

8.3 This is done by squaring a typical OR element. For G_1^3,

$$(ra^i b^j)^2 = r^2 (a^i b^j)^r (a^i b^j)$$
$$= a(a^i b^{-j})(a^i b^j)$$
$$= a^{2i+1} \neq 1.$$

8.4 Of the four groups, only G_1 is abelian. Of the remaining three, only G_1^3 has no element of order 2. For the other two, we reduce the relation matrices to Smith normal form:

$$\begin{bmatrix} 0 & 0 & 0 \\ 0 & 0 & 2 \\ 0 & 0 & 0 \\ 0 & 2 & 0 \end{bmatrix} \sim \begin{bmatrix} 2 & 0 & 0 \\ 0 & 2 & 0 \\ 0 & 0 & 0 \\ 0 & 0 & 0 \end{bmatrix}, \quad \begin{bmatrix} 0 & 0 & 0 \\ 0 & 0 & 2 \\ 1 & -1 & 0 \\ -1 & 1 & 0 \end{bmatrix} \sim \begin{bmatrix} 1 & 0 & 0 \\ 0 & 2 & 0 \\ 0 & 0 & 0 \\ 0 & 0 & 0 \end{bmatrix}.$$

The derived factor groups are thus $Z_2 \times Z_2 \times Z$ and $Z_2 \times Z$ respectively, whence G_1^1 and G_1^2 are not isomorphic.

8.7 Since $a^s = b$ and $r^2 = a$, $G_4^2 = \langle r, s \rangle$. Using Tietze transformations to eliminate $a = r^2$, $b = (r^2)^s$, the original eight relations become $r^2 (r^2)^s = (r^2)^s r^2$, $s^4 = 1$, $-$, $(r^2)^{s^2} = r^{-2}$, $-$, $(r^2)^r = r^2$, $(r^2)^{sr} = (r^{-2})^s$, $(sr)^2 = 1$. At least three of these six are redundant, and so

$$G_4^2 = \langle r, s \mid s^4 = (sr)^2 = 1, (r^2)^{srs^{-1}} = r^{-2} \rangle.$$

8.8 Faute de mieux, we bulldoze our way through this one by computing all centres of rotation and all axes of reflection in both cases, working with normal forms and omitting routine calculations. Referring to Fig. 9.3 in Chapter 9, we proceed in two steps.

Step 1. The rotations in both groups are the same, namely, the elements $s^{\pm 1} a^k b^l$, where $s = s(O, 2\pi/3)$, a and b are translations sending O to A, B respectively, and $k, l \in \mathbb{Z}$. To find their centres, we put the conjugate s^g, where $g = a^i b^j$, into normal form:

$$s^g = g^{-1} s g = a^{-i} b^{-j} s a^i b^j$$
$$= s s^{-1} a^{-i} b^{-j} s a^i b^j$$
$$= s (a^s)^{-i} (b^s)^{-j} a^i b^j$$
$$= s (a^{-1} b)^{-i} (a^{-1})^{-j} a^i b^j$$
$$= s a^i b^{-i} a^j a^i b^j$$
$$= s a^{2i+j} b^{-i+j}.$$

These are the rotations $sa^k b^l$ for which $3 \mid (k-l)$, and their centres are all points of the form $Og = Oa^i b^j$ by the result of Exercise 6.1, that is, the lattice points of the regular tessellation $\mathcal{T}$ of $\mathbb{R}^2$ generated by the triangle OAB. If $k-l \equiv \pm 1 \pmod 3$, we apply the same reasoning to the rotations $sa^{\pm 1}$, whose centres are the points P, Q shown in Fig. A(i), where P is the centre of the triangle OAB and $Q = -P$. The images of P, Q under all $g = a^i b^j$ are the centres of all triangles in $\mathcal{T}$. Since sg and $(sg)^{-1}$ have the same centre, we have found all of them. They are the vertices and centres of all triangles in $\mathcal{T}$.

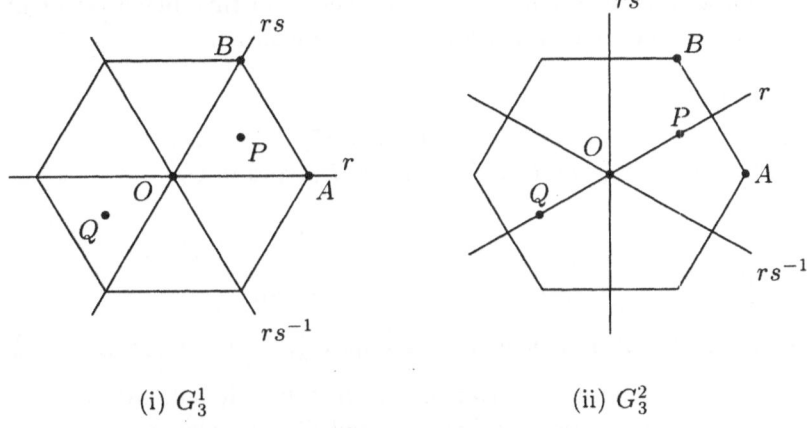

(i) G_3^1 (ii) G_3^2

Figure A   Axes of reflection in G_3-groups

Step 2. To find the axes of reflection in G_3^1, first note that the axes of r, rs, rs^{-1} are as indicated in Fig. A(i), namely, the lines through O and A, O and B, O and Br respectively. All reflections are found by putting the square of a typical OR element $rs^k a^i b^j$ equal to 1, $i, j, k \in \mathbb{Z}$, $0 \leq k \leq 2$. They are then expressed as conjugates of rs^k. When $k = 0$ we get

$$(ra^i b^j)^2 = (a^r)^i (b^r)^j a^i b^j$$
$$= a^i (ab^{-1})^j a^i b^j$$
$$= a^{2i+j} = 1 \Leftrightarrow j = -2i,$$

and then find that $ra^i b^{-2i} = r^{b^{-i}}$. The axes of these reflections are thus the images under powers of b of the axis of r, namely, all the horizontal lines in $\mathcal{T}$. When $k = \pm 1$, we can either proceed in the same way or simply conjugate the whole business by $s^{\pm 1}$. The result is that the axes of reflection in G_3^1 are just the lines in $\mathcal{T}$.

In the case of G_3^2, the axes of the reflection of rs^k, $0 \leq k \leq 2$, are as marked in Fig. A(ii). As above, we show that

$$(ra^i b^j)^2 = 1 \Leftrightarrow j = -i,$$

and then $ra^i b^{-i} = r^{a^i}$. The axes of these reflections are thus obtained by applying powers of a to the axis of r. Applying $s^{\pm 1}$ to these axes gives the corresponding result for reflections of the form $rs^{\pm 1} a^i b^j$, namely, $rs^{-1} b^j = (rs^{-1})^{a^j}$ and $rsa^i = (rs)^{b^i}$. So the axes here are the perpendicular bisectors of pairs of adjacent vertices of $\mathcal{T}$.

Thus the axes in G_3^2 contain the vertices and centres of all triangles in $\mathcal{T}$, but those in G_3^1 contain only the vertices. And this, believe it or not, is the easiest way to distinguish between these two groups.

8.9 It is G_3.

8.10 In G_1^1 and G_1^2, $C = \langle r \rangle \cong Z_2$, in G_2^1 and G_2^3, $C = \langle r, s \rangle \cong Z_2^2$, in G_4^1, $C = \langle r, s \rangle \cong D_5$, in G_3^1 and G_3^2, $C = \langle r, s \rangle \cong D_6$, and in G_6^1, $C = \langle r, s \rangle \cong D_{12}$.

Chapter 9

9.2-9.9 Figs. 9.19–9.26 have symmetry groups G_2^3, G_2, G_2, G_2^4, G_2^2, G_1^3, G_6^1, G_3^1.

9.10 Such a tessellation is given in the text for all but G_1, G_1^3, G_6^1, and for G_6^1 we have Fig. 9.25. For G_1 and G_1^3, the diagonals in Figs. 9.13 and 9.14 can be rotated about their centres through $\pm \varepsilon$ (according to orientation) to give tessellations by congruent trapezia.

Chapter 10

10.1 (a) A circle, (b) a square.

10.2 (a) False, because of the spherical excess.

(b) True, using isometries.

(c) True by default: there are no parallelograms.

(d) False: an octant has three right angles and three equal sides.

10.3 See Guide to the Literature.

10.7 From Eqs. (10.4) and (10.5),

$$v = fn/m, \quad e = fn/2, \quad v - e + f = 2,$$

get f in terms of m, n by eliminating v, e:

$$f(n/m - n/2 + 1) = 2$$
$$\Rightarrow \ f(2n - mn + 2m) = 4m,$$

and divide by $2n + 2m - mn = 4 - (m - 2)(n - 2)$.

10.8 $(2, m)$ consists of m points equally spaced round the equator and $(m, 2)$ is its dual, made up of m congruent lunes concurrent at the poles. For v, l, f we get m, m, 2 and 2, m, m respectively, and the group is D_{2m} in both cases.

10.9 By the addition formula of trigonometry,

$$2 \sin \theta \cos \theta = \sin 2\theta = \cos 3\theta = \cos 2\theta \cos \theta - \sin 2\theta \sin \theta$$
$$= (1 - 2 \sin^2 \theta) \cos \theta - 2 \sin^2 \theta \cos \theta.$$

Putting $s = \sin \theta$ and cancelling $\cos \theta$,

$$4s^2 + 2s - 1 = 0,$$

whence $s = \frac{\sqrt{5}-1}{4} = \sin \pi/10$. Then

$$2 \cos \pi/5 = 2(1 - 2s^2) = 2 \left(1 - \frac{3 - \sqrt{5}}{4} \right) = \frac{1 + \sqrt{5}}{2}.$$

Solving Eq. (10.7), we also get $\tau = \frac{1+\sqrt{5}}{2}$.

10.11 The point $P = (\tau, 1, 0)$ lies at distance 2 from each of

$$(0, \tau, 1), \ (1, 0, \tau), \ (\tau, -1, 0), \ (1, 0, -\tau), \ (0, \tau, -1).$$

Labelling these points Q_i, $1 \le i \le 5$, in this cyclic order, each lies at distance 2 from the next. Likewise with $-P$ and the $-Q_i$, $5 \ge i \ge 1$. Q_1 also lies at distance 2 from $-Q_3$ and $-Q_4$, and cyclically. So we get a regular icosahedron of radius

$$\frac{1}{2} d(P, -P) = \frac{1}{2} \sqrt{4\tau^2 + 4} = \sqrt{\tau + 2}.$$

10.13 We have $\mathrm{Sym}^+(\mathbf{C}) \cong S_4$. Since $\mathrm{Sym}(\mathbf{C})$ contains reflection $r : (x, y, z) \mapsto (-x, -y, -z)$ in the origin and r is central, OR and of order 2, it follows that $\mathrm{Sym}(\mathbf{C}) = \mathrm{Sym}^+(\mathbf{C}) \times \langle r \rangle \cong S_4 \times Z_2$.

10.16 We want the angle 2θ between the lines joining the midpoint of an edge to the diametrically opposite points of the two faces that contain it. Using the coordinates given in the text and Exercises 10.12 and 10.11 above for the vertices of $\mathbf{T}$, $\mathbf{C}$, $\mathbf{O}$, $\mathbf{D}$, $\mathbf{I}$, we get the value of $\sin \theta$ to be

$$1/\sqrt{3}, \ 1/\sqrt{2}, \ \sqrt{2/3}, \ (2 + \tau)/5, \ \tau/\sqrt{3},$$

respectively.

Chapter 11

11.2 That $|G : H| = 2$ is also a consequence of the next exercise.

11.3 The automorphism y sends the relations (11.2) to

$$a^{-l} = b^{-m} = (ba)^{-n} = 1.$$

Inverting each of these and conjugating the last by b yields (11.2). The resulting semidirect product $\langle y \mid y^2 = 1 \rangle \times_\alpha H$ has generators a, b, y and defining relations

$$a^l = b^m = (ab)^n = 1, \ y^2 = 1, \ a^y = a^{-1}, \ b^y = b^{-1}.$$

In the presence of $y^2 = 1$, the last two relations can be rewritten $(ay)^2 = 1$, $(yb)^2 = 1$. In terms of the new generators $x = ay$, $y = y$, $z = yb$ ($a = xy$, $b = yz$), we have defining relations (using (5.14))

$$(xy)^l = (yz)^l = (xy^2z)^n = y^2 = x^2 = z^2 = 1$$

equivalent to those of $\Delta(l, m, n)$ in (11.1).

11.4 From the relation matrix (see Section 5.5) we see that $\Delta(l, m, n)$ has derived factor group Z_2^3, Z_2^2, Z_2 according as the number of l, m, n that are even is $3, 2, \leq 1$.

For $\Delta^+(l, m, n)$ the rows of the relation matrix are $(l, 0)$, $(m, 0)$, (n, n). A little thought shows that elementary row operations do not change the hcf h of the entries l, m, n nor the hcf k of the 2×2 subdeterminants lm, ln, mn. It then follows from the Smith normal form that the derived factor group of this group is $Z_h \times Z_{k/h}$.

11.7 Using the fact that $t = as$ and a process of substitution, (5.14) yields defining relations

$$s^3 = (as)^3 = (sas)^3 = 1, \ s = s, \ a = a, \ b = as^2as.$$

Working from these, we have $s^3 = 1$, whence $b = as^{-1}as$, that is, $a^s = a^{-1}b$. Next, $1 = sas^2as^2as = sbsas$, so that

$$b^s = s^2bs = ss^{-1}a^{-1} = a^{-1}.$$

Finally,

$$b^a = (as^2as)^a = s^2asa = a^sa = b,$$

so that $ab = ba$ and we have deduced the relations of G_3. The reverse implication is similar but simpler.

11.8 No.

11.9 Use Exercises 11.5, 11.6 and 11.7 and take the translation subgroup in each case.

11.10 (b) The dual of the graph in (a), labelled as in (c), looks like this.

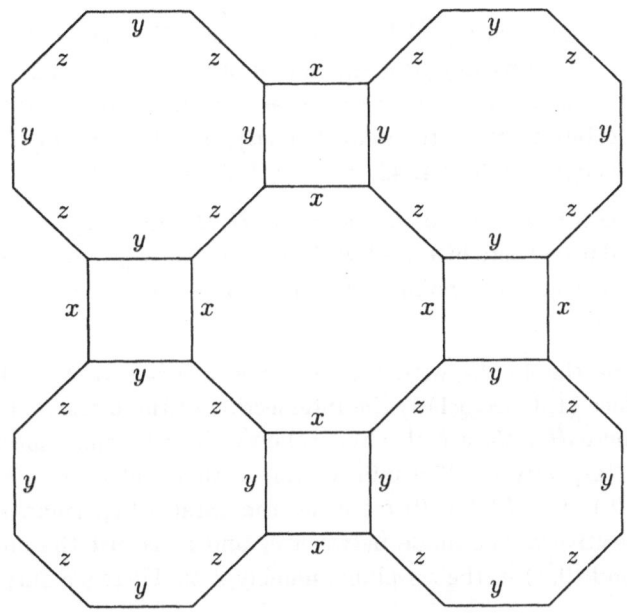

Figure B   Illustrating the proof of Theorem 11.1

$$w = r^{zy}s^{xyzy}r^{xzxyzy}t^{xyzy}r^{xzxyzy}r^{zyzy}s.$$

11.11 For $Z_8 = \langle x \mid x^8 \rangle$, label every edge x and orient them all anticlockwise. For $D_8 = \langle x, y \mid x^2, y^2, (xy)^4 \rangle$, label the (unoriented) edges x, y alternately.

11.12 For Z_2^3 and $Z_2 \times Z_4$ you should get (suitably labelled and oriented) concentric squares with each of four pairs of corresponding vertices joined by an edge. The Cayley diagram of Q_8 cannot be embedded in $\mathbb{R}^2$.

11.15 $\langle x, y, z \mid x^2 = y^2 = z^2 = 1 \rangle$.

11.18 At least three: $G_4^1 = \langle r, sr, br \rangle$, $G_3^2 = \langle r, sr, ab^{-1}r \rangle$, $G_6^1 = \langle r, sr, ab^{-1}r \rangle$.

11.19 $\Delta(2, 2, 2) \cong A_1^3$, $\Delta(2, 2, n) \cong A_1 \times I_2(n)$, $\Delta(2, 3, 3) \cong A_3$, $\Delta(2, 3, 4) \cong A_1 \times A_3$, $\Delta(2, 3, 5) \cong H_3$. (Here A_n refers to the group so labelled in Fig.

11.5, and not to the alternating group of degree n.) Also

$$\Delta(3,3,3) \cong \tilde{A}_2, \quad \Delta(2,4,4) \cong \tilde{B}_2, \quad \Delta(2,3,6) \cong \tilde{G}_2.$$

Chapter 12

12.3 Let P be any polytope with symmetry group G and set Fl of flags. Since only $1 \in G$ can fix any flag, the action of G on Fl is regular if and only if it is transitive, and this in turn is equivalent to the equality $|G| = |Fl|$ by the orbit-stabiliser theorem. For $\{n\}$, $n \geq 3$, **T**, **C**, **O**, **D**, **I**, $|G|$ and $|Fl|$ are equal, to $2n$, 24, 48, 48, 120, 120 respectively.

12.5 Let ϕ be a flag of H so that $\phi_i = (\phi, H, H_i)$ is a flag of P for $i = 1, 2$. By regularity, there is a $g \in \text{Sym}(P)$ such that $\phi_1 g = \phi_2$. Since g fixes H and ϕ, it fixes every point of H. Since $H_1 g = H_2$ also, g is the required reflection.

12.6 Let P be the 4-cube with vertices $(x, y, z, t) = (\pm 1, \pm 1, \pm 1, \pm 1)$. Then the 2-face $(1, 1, \pm 1, \pm 1)$ is the intersection of the faces $P \cap H_1$ and $P \cap H_2$, where H_1, H_2 are the hyperplanes given by the equations $x = 1$, $y = 1$ respectively. The unit vectors orthogonal to H_1, H_2 are $v_1 = (1, 0, 0, 0)$, $v_2 = (0, 1, 0, 0)$ (consider the parallel hyperplanes $x = 0$, $y = 0$) respectively. The angle between v_1 and v_2 is just the angle between $(1, 0)$ and $(0, 1)$ in the xy-plane, namely $\pi/2$. There's a surprise.

12.8 The vertices of $\mathbf{T}_d$ are the points P_i of H with ith coordinate -1 and the other d equal to 1, $1 \leq i \leq d + 1$. Then all but P_i lie in the hyperplane H_i in $\mathbb{R}^{d+1}$ with equation $x_i = 1$, $1 \leq i \leq d + 1$. So $\mathbf{T}_d$ is defined by the hyperplanes $H_i \cap H$ in H, $1 \leq i \leq d + 1$.

12.10 Its Schläfli symbol $\{3, 4, 3\}$ is symmetric.

12.11 The vertices $(\tau, \pm 1, \pm \tau^{-1}, 0)$, $(\tau, 0, \pm 1, \pm \tau^{-1})$, $(\tau, \pm \tau^{-1}, 0, \pm 1)$ of the given link P all lie in the hyperplane $v = \tau$ in $vxyz$-space $\mathbb{R}^4$, and hence so does P. Dropping the v-coordinate and multiplying the others by τ, we have twelve vertices

$$(\pm \tau, \pm 1, 0), \quad (0, \pm \tau, \pm 1), \quad (\pm 1, 0, \pm \tau)$$

of an icosahedron in $\mathbb{R}^3$ (see Exercise 10.11).

12.12 Let **P** be the 600-cell with centre O and vertex set as given in the text, and **P′** its link (an icosahedron by the previous exercise) at the vertex $v = (2, 0, 0, 0)$, with vertices

$$(\tau, \pm 1, \pm \tau^{-1}, 0), \quad (\tau, 0, \pm 1, \pm \tau^{-1}), \quad (\tau, \pm \tau^{-1}, 0, \pm 1).$$

Then a flag of $\mathbf{P}$ is obtained from v by successively adjoining the vertices

$$v_1 = (\tau, 0, 1, \tau^{-1}), \quad v_2 = (\tau, 1, \tau^{-1}, 0), \quad v_3 = (\tau, \tau^{-1}, 0, 1).$$

The triangular face $v_1 v_2 v_3$ of $\mathbf{P}'$ is represented, along with its three neighbours, in Fig. C by specifying the last three coordinates of each vertex (the first is $v = \tau$). Then the four tetrahedra fomed from these triangles by adjoining the vertex v are faces of $\mathbf{P}$, whose centres X, P, Q, R are vertices of $\mathbf{P}^*$, with coordinates

$$\frac{1}{4}(\tau^4, \tau, \tau, \tau), \quad \frac{1}{4}(\tau^4, 1, \tau^2, 0), \quad \frac{1}{4}(\tau^4, \tau^2, 0, 1), \quad \frac{1}{4}(\tau^4, 0, 1, \tau^2),$$

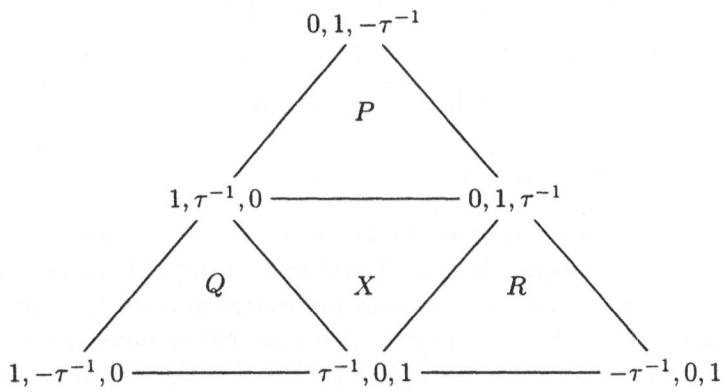

Figure C   Four faces of the icosahedron $\mathbf{P}'$

respectively. Thus P, Q, R are three vertices of the link of $\mathbf{P}^*$ at X, and it remains to find the fourth, call it S. Since the centre of this tetrahedron lies on the line through O and X, we can take $S = \frac{1}{4}(a, b, b, b)$, where

$$3\tau^4 + a = \lambda \tau^4, \quad 1 + \tau^2 + b = \lambda \tau.$$

But since P is equidistant from S and Q, we have

$$\tau^4 - \frac{1}{8}(a\tau^4 + b + b\tau^2) = \tau^2/4.$$

Eliminating a and b from these three equations, we get

$$(\lambda - 3)\tau^8 + (\lambda\tau - 2 - \tau)(1 + \tau^2) = 8\tau^4 - 2\tau^2,$$

giving $\lambda = 3\tau - 1$ and $S = \frac{1}{4}(3\tau + 1, 1 + \tau, 1 + \tau, 1 + \tau)$.

12.14 T_d has $d+1$ vertices, any $k+1$ of which form the vertex set of a T_k. So the number of k-faces is

$$f_k(T_d) = \binom{d+1}{k+1}, \quad 0 \le k \le d-1.$$

A k-face of C_d has vertex set obtained by allowing ± 1 in each of k chosen coordinate positions and fixing the entry in each of the other $d-k$. These k positions can be chosen in $\binom{d}{k}$ ways and the fixed entries in the others in 2^{d-k} ways. So $f_k(C_d) = \binom{d}{k} 2^{d-k}$, $0 \le k \le d-1$.

By duality, $f_k(C_d^*) = f_{d-1-k}(C_d) = \binom{d}{k+1} 2^{k+1}$, $0 \le k \le d-1$.

In each case, a calculation using the binomial theorem shows that $\sum_{k=0}^{d-1} (-1)^k f_k = 1-(-1)^d$. Reasonably putting $f_d = 1$ (the polytope itself), we guess $\sum_{k=0}^{d} (-1)^k f_k = 1$.

12.16 When $d = 2$ there are three: $(4,4)$, $(3,6)$, $(6,3)$, as described in Chapter 9. When $d = 3$ there is just one, by cubes, as stated in Exercise 10.17. The hint confirms this and gives the analogous tessellations, by d-cubes, in all dimensions $d \ge 4$. It also proves, after some calculations, the uniqueness of this tessellation for any $d \ge 5$ (it is self-dual) and the existence of exactly two more when $d = 4$: by copies of $C_4^* = \{3,3,4\}$ and its dual by copies of the 24-cell $\{3,4,3\}$, with $r_4 = 3$ in each case.

12.17 The values shown are easily computed using Theorem 12.1 and the fact that $\rho\{n\} = \sin^2 \pi/n$. Easy inductions then show that

$$\rho(T_d) = (d+1)/2d, \quad \rho(C_d) = 1/(d+1), \quad \rho(C_d^*) = 1/2.$$

Guide to the Literature

There follows a description of sources, alternative treatments and further reading organised on a chapter-by-chapter basis. Of course the literature on symmetry is truly vast, and the skimpy list of 19 items below is meant merely as a small but fairly representative selection. It is hoped that nearly all are accessible in both senses: they are obtainable from many bookshops and libraries, and are capable of being understood by most normal human beings.

1. Both logically and chronologically, the theory of metric spaces forms a bridge between analysis and topology. A thorough treatment in this context is to be found in [17].

Good introductions to group theory *per se* are to be found in the books [11] and [18], the latter in this series. The justly popular classic [19] forms a readable account of symmetry from many perspectives.

2. Most of the material on the Euclidean group enjoys the status of folklore. The treatment here follows that in the wonderfully surrealistic [12], which in large measure inspired this book.

3. The theory of groups is taken further in the excellent books [9] and [16], the latter in this series. In both cases the standard of exposition is extremely high.

4. I made most of this up as I went along.

5. The last chapter of [16] forms a clear introduction to the theory of group presentations, and [10] is a fairly standard text containing references to more advanced studies.

6, 7, 8. These chapters form the core of the book and follow [12] pretty closely, and we use the notation in that book for both friezes and plane crystallographic groups. The many notations for the latter are carefully tabulated in the aptly titled article [15], from which we borrowed the means described in Exercise 8.8 of distinguishing between the notoriously similar groups G_3^1 and G_3^2.

9. It is to be hoped that there is a splash of novelty in our diagrams. They were inspired by an exercise in [12], which contains an alternative set of 17 figures. In fact there are many such sets, of which perhaps the best known is that in [6].

10. I got Girard's theorem from Ronnie Brown, the construction of **D** from John Wilson, and the coordinates of **I** from [7]. The platonic solids appear in many of the references below and are derived by an entirely different method in the Appendix of [19]. A solution to the rather difficult Exercise 10.3 is to be found in [4].

11. Triangle groups have formed an object of interest for more than 100 years, and were among the first groups to be studied in terms of a presentation by generators and relations. A very good introduction to the general theory of Coxeter groups is to be found in [1], which also contains a complete and accessible derivation of the groups tabulated in Figs. 11.5 and 11.6. More advanced coverage is given in the wide-ranging survey [8] and the seminal Chapter IV of [3]. Incidentally, this last reference is probably the most frequently cited of all those in our list and is certainly the only one not in English.

12. The glorious topic of this chapter is mentioned in [8], [12] and elsewhere. Fuller treatments appear in [5] and [13], which are in and out of print respectively. The approach here follows that in [2], which, like [4] and [14], gives a more encyclopedic coverage of geometry than that attempted here. All three of these books are highly recommended for further reading.

Bibliography

[1] C. T. Benson and L. C. Grove, *Finite Reflection Groups*, Bogden-Quigley, Tarrytown, NY, 1971.

[2] M. Berger, *Geometry* I, II, Springer-Universitext, Paris, 1992, 1996.

[3] N. Bourbaki, *Groupes et Algèbres de Lie*, Ch. IV–VI, Hermann, Paris, 1960.

[4] D. A. Brannan, M. F. Esplen and J. J. Gray, *Geometry*, Cambridge University Press, 1999.

[5] H. S. M. Coxeter, *Regular Polytopes*, 3rd ed., Dover, NY, 1973.

[6] L. Fejes Tóth, *Regular Figures*, Macmillan, NY, 1964.

[7] J. Goethals and J. J. Seidel, The football, *Nieuw Archief voor Wiskunde* **29** (1981), 50–58.

[8] P. de la Harpe, An invitation to Coxeter groups. In *Group Theory from a Geometrical Viewpoint*, World Scientific, Singapore, 1991, pp. 193–253.

[9] J. F. Humphreys, *A Course in Group Theory*, Oxford University Press, Oxford, 1996.

[10] D. L. Johnson, *Presentations of Groups*, 2nd ed., Cambridge University Press, Cambridge, 1997.

[11] W. Ledermann and A. J. Weir, *Introduction to Group Theory*, 2nd ed., Addison-Wesley, Harlow, 1996.

[12] R. C. Lyndon, *Groups and Geometry*, Cambridge University Press, Cambridge, 1985.

[13] H. P. Manning, *Geometry of Four Dimensions*, Macmillan, NY, 1914.

[14] P. M. Neumann, G. A. Stoy and E. C. Thompson, *Groups and Geometry*, Oxford University Press, Oxford, 1990.

[15] D. Schattschneider, The plane symmetry groups: their recognition and notation, *Amer. Math. Monthly* **85** (1978), 439–450.

[16] G. C. Smith and O. A. Tabachnikova, *Topics in Group Theory*, SUMS, Springer-Verlag, London, 2000.

[17] W. A. Sutherland, *Introduction to Metric and Topological Spaces*, Clarendon, Oxford, 1998.

[18] D. A. R. Wallace, *Groups, Rings and Fields*, SUMS, Springer-Verlag, London, 1998.

[19] H. Weyl, *Symmetry*, Springer-PUP, Princeton, 1952.

Index of Notation

The use of symbols in mathematics is of fundamental importance. Not only do they provide convenient abbreviations, but correct choice of notation can often play a crucial role in solving a problem. And conversely: if you don't believe me, try dividing CCXXX by XVII. On the other hand, many ways of abusing notation spring readily to mind. Two of these are, for historical reasons, especially popular. First, different symbols may be used to represent the same quantity or concept. A fine example is the semidirect product of groups: it is no easy matter to come up with a symbol that has not been used previously. Second, and not quite so harmless, is the use of the same symbol to stand for different objects. Thus the symbols A_n is used both for the alternating group of degree n and the symmetric group of degree $n + 1$. It is only to be hoped that such ambiguity is resolved by appealing to the context.

Dramatis Personae (in Approximate Order of Appearance)

SSS, SAS, SAA, RHS	sufficient conditions for the congruence of triangles
d	(usually) a metric
$\mathbb{R}$	the set of real numbers, the real line
$\forall$	universal quantification (for all)
(X, d)	a metric space
$\mathbb{R}^2$	the Euclidean (or Cartesian) plane
x, **y**, **z**	vectors
$\mathrm{Isom}(X, d)$, $\mathrm{Isom}(X)$	group of isometries
S_n	the symmetric group of degree n

$n!$	n factorial
$\mathbb{E}$	the Euclidean group
G	a group
t, s, r, q	(usually) a translation, rotation, reflection, glide reflection
$\mathbb{N}$	positive integers
OP	orientation preserving, order preserving
OR	orientation reversing, order reversing
G^+	OP subgroup of G
G^-	set of OR elements of G
$x^y = y^{-1}xy$	the conjugate of x by y
$[x, y] = x^{-1}y^{-1}xy$	the commutator of x and y
$\mathrm{Sym}(F)$	the symmetry group of F
$\mathbb{Z}$	the set of integers
D_∞	the infinite dihedral group
xH	left coset of subgroup H containing $x \in G$
G'	the derived group (or commutator subgroup) of a group G
$\mathbb{Q}$	the set of rational numbers
$A\hat{B}C$	angle between BA and BC
π	pi, the ratio of the circumference of a circle to its diameter
$\cap$	intersection of sets
$\cup$	union of sets
$\emptyset$	the empty set
$\mathbb{T}$	translation subgroup
$\mathbb{S}_0$	subgroup of rotations with centre O
$\mathbb{C}$	the complex numbers
$\bar{z}$	complex conjugate of $z \in \mathbb{C}$
$\mathbb{E}^+$	the subgroup of OP isometries of $\mathbb{R}^2$
$|G|$	order (cardinality) of a group G
$\cong$	isomorphism (of groups)
$\mathrm{Aut}(G)$	automorphism group of a group G
$\mathrm{Inn}(G)$	group of inner automorphisms
$\mathrm{Im}\,\phi$	image of homomorphism ϕ
$\mathrm{Ker}\,\phi$	kernel of ϕ
$Z(G)$	centre of G
$\langle X \rangle$	subgroup generated by X
$X^\pm$	elements of X and their inverses
Z	the infinite cyclic group
Z_n	the cyclic group of order n

$\lvert x \rvert$	order of element x, modulus of number x
$\mathbf{0}$	origin of coordinates (usually) in $\mathbb{R}^2$
Hx	right coset of H containing x
$H \lhd G$	H is a normal subgroup of G
G/H	factor group of G by H
G/G', G^{ab}	derived factor group of G, G abelianised
$K \times H$	Cartesian, or direct, product
$K \times_\alpha H$	semidirect product of H and K
	with respect to α
D_{2n}	dihedral group of order $2n$
$\mathbb{Z}^2$	the set of integer points in the plane
Z^2	free abelian group of rank 2
aG	orbit of a under the action of G
$\mathrm{Stab}_G(a)$	stabiliser of a in G
$\overrightarrow{PP'}$	vector from P to P', $P, P' \in \mathbb{R}^2$
$F(X)$	free group on X
$\langle X \mid R \rangle$	group presentation
$\overline{R}$	normal closure of R
Z^n	free abelian group of rank n
$(a_1 a_2 \cdots a_l)$	cyclic permutation
A_n	(usually) alternating group of degree n
$\Delta(l, m, n)$	triangle group
$\mathbb{S}^2$	the sphere
$\mathbb{H}^2$	the hyperbolic plane
$\tau(G)$	torsion subgroup of G
(k, l)	(sometimes) highest common factor of integers k, l
Ω_n	the set of the first n positive integers, $n \in \mathbb{N}$
$\exists$	the existential quantifier (there is)
$\mathbf{P}, \mathbf{M}, \mathbf{A}$	elementary (row) operations
F_i^j	frieze groups
G_i^j	plane crystallographic groups
$(4, 4), (3, 6), (6, 3)$	regular tessellations of the plane
$\mathbf{T}$	tetrehedron
$\mathbf{C}$	cube
$\mathbf{O}$	octahedron
$\mathbf{D}$	dodecahedron
$\mathbf{I}$	icosahedron
Δ	triangle
τ	the golden section
Sym	symmetry group
Sym^+	group of OP symmetries

AP	Axiom of Parallels
$\mathbb{R}^n$	Euclidean n-space
$\mathbb{P}^n$	elliptic n-space
$\mathbb{H}^n$	hyperbolic n-space
$GL(n, \mathbb{R})$	general linear group
K_n	complete graph on n vertices
$\mathbf{C}_d$	d-cube
$\mathbf{T}_d$	regular d-simplex
$\mathbf{C}_d^*$	d-cocube
$f'(x)$	derivative of function $f(x)$
$\parallel$	is parallel to
$\perp$	is perpendicular to
$\{r_1, r_2, \ldots, r_{d-1}\}$	Schläfli symbol of a d-dimensional regular polytope
$\binom{d}{k}$	binomial coefficient, d choose k

Index